Cave and Karst Systems of the World

Series Editor

James W. LaMoreaux, P.E.LaMoreaux and Associates, Tuscaloosa, AL, USA

More information about this series at https://link.springer.com/bookseries/11987

Gültekin Günay · Koray Törk ·
İsmail Noyan GÜNER ·
Eric Gilli

Caves and Karst of Turkey—Volume 2

Geology, Hydrogeology and Karst

Springer

Gültekin Günay
Hydrogeology Engineering
Faculty of Engineering
Hacettepe University, Beytepe Campus
Ankara, Turkey

İsmail Noyan GÜNER
Energy Raw Material Research and Exploration
Mineral Research and Exploration of Turkey
Ankara, Turkey

Koray Törk
Geological Research
Mineral Research and Exploration of Turkey
Ankara, Turkey

Eric Gilli
Paris 8 University
Saint-Denis, France

ISSN 2364-4591 ISSN 2364-4605 (electronic)
Cave and Karst Systems of the World
ISBN 978-3-030-95360-7 ISBN 978-3-030-95361-4 (eBook)
https://doi.org/10.1007/978-3-030-95361-4

This Springer imprint is published by the registered company Springer Nature Switzerland AG
The registered company address is: Gewerbestrasse 11, 6330 Cham, Switzerland

Dumanli Karst Springs (Underground River)—The Largest Karst Spring in the World—Manavgat River, Antalya, Turkey.The Dumanli Spring in the Mediterranean region of Turkey was already submerged by the year 1982 by about 120 m head produced by the Oymapinar reservoir. Manavgat River will be dammed at Oymapinar. The mean discharge of the spring is estimated at about 50 m^3/s; in October 1978, the authors measured the spring discharge by the Dye-Dilution technique. The flow rate of about 35.6 m^3/s at the very and of the dry period (at the end of spring's discharge recession) motivated the authors to declare the Dumanli, the largest karst spring in the world, issuing from one single orifice (G. Günay, J. Karanjac—*Journal of Hydrology*, 1980).

Foreword

Between 1977 and 2000, a series of six symposiums were convened in Turkey. These symposiums are considered to be the most important reunions of the world's karst society. The main organizer and the initiator of these worldwide known karst conferences was Prof. Dr. Gültekin Günay from Hacettepe University, Ankara, Turkey.

Günay's vast knowledge and experience in geology and particularly in karstology are based on many years of research and field visits in various karst regions of Turkey. His in-depth knowledge in engineering has been an important asset while he was serving as the leading geologist of the General Directorate of State Hydraulic Works of the Republic of Turkey (DSİ). Under the umbrella of the United Nations Development Program (UNDP), Dr. Günay established the International Research and Application Centre for Karst Water Resources (UKAM) at Hacettepe University. In a short period, UKAM has become the leading institution for karstology in Turkey and one of the top-ranking karst research centers in the world. Under the supervision of Dr. Günay, UKAM has progressed toward being one of the leading schools of well-educated and trained young specialists in the field of scientific and applied karstology. For many years, Günay has been a member of an important Commission of the International Association of Hydrogeologists—the Karst Commission. He was the chairman of all six Karst symposiums (1977, 1979, 1985, 1990, 1995, 2000) and editor or co-editor of all symposium proceedings. Based on his extraordinary experiences, he published many scientific papers in various journals and submitted them to symposiums. In a great number of papers, as an author and co-author, he analyzed and presented the geological and hydrogeological features of numerous karst regions of Turkey, particularly the Taurus karst region, which is the largest and the most significant karst region in Turkey.

With more than 40 years of experience in the field, Dr. Günay has been involved in an extensive number of projects in many karst regions in Turkey. In this book, Günay shares the vast experience that he has accumulated during his professional life. The content of the book mainly includes his fieldwork, application of various investigation methods, data analysis, and presentation in the form of diverse professional and scientific papers. It is also based on the undergraduate and master courses in karst hydrogeology taught at Hacettepe University.

This book consists of two volumes. Each volume encompasses distinct topics focused on specific karst phenomena on a regional scale and a detailed explanation of many case studies. In the first part, hydrogeological characteristics of important karst river basins, including specific karst phenomena such as sinkholes (obruks), huge deposits of travertine in the Antalya region, gypsum karst of the Sivas region, and particular groundwater flow systems, are depicted. One of the essential characteristics of Turkish karst is fascinating karstic springs. Attributes of these springs are presented in detail. Some of them are well known to anybody dealing with karst hydrogeology: Dumanli Spring, one of the largest karst springs in the world; Pamukkale, a captivating geothermal spring; Kirkgözler Spring in Antalya travertine area and a few others. The text includes the results of remarkable studies performed for the protection of water quality of karstic springs and geothermal reservoirs.

This book contains a vast knowledge of the general characteristics of karst regions in Turkey. It is a valuable source for scientists and engineers working in these regions and a significant handbook for other professionals working in numerous karst regions of the world. One of the aims of this book is to assist and encourage students of the earth sciences and engineering to grasp karst aquifer properties from a technical and environmental point of view.

It is an extraordinary book with precious data for planners of regional socio-economic development in such an unfriendly environment as karst landscapes.

Belgrade, Serbia

Prof. Dr. Petar Milanovic
Emeritus Professor

Preface

Karstic limestones in Turkey cover approximately one-third of the country and are primarily located in the Mediterranean sector of the Alpine orogenic belt. This belt lies in between the Russian belt to the north and the African and Arabic belts to the south. The geologic history of the region was influenced by the repetitive opening and closing of several ocean basins, whose tracers are depicted as suture belts. Karst limestone areas in Turkey, covering one-third area, also include very productive aquifers. The karstification in limestone is more common in Paleozoic, Cretaceous, Miocene, and Neogene formations. The carbonate rock units are about 200 km wide along the Taurus Mountains, which attain elevations of 2500 m. These high mountain ranges, sharp peaks, deep valleys, and narrow gorges cause extremely rugged topography. The Taurus Mountains were formed by folding and overthrust faulting during the Alpine orogeny. With uplift, karstification increased and the surface drainage of many lakes and rivers was transformed into subsurface drainage. Turkey has several types of karstic landforms containing lapies, caverns, dolines, uvalas, poljes, and underground river valleys. Karstification is related not only to the thickness and purity of limestone, climate, and height but also to tectonic movements. Well-developed karstic features such as poljes, groundwater, and cave systems are widespread in/on the Mesozoic comprehensive limestone in the Taurus Mountains. The formation of these karstic forms has taken a long time. Karstification has begun to be developed toward the end of the Mesozoic limestones by the uplifting movements of the Taurus Mountains in general. Tectonic movements are considerably responsible for karstification and karstic landforms. Most of the deep and large karstic depressions, especially poljes, are related to vertical tectonic movements. The Taurus Mountain region is characterized by abundant water resources, large hydroelectric potential, some of the world's largest karst aquifers, and the largest karst springs.

These springs of exceedingly large discharge are formed by the channelized flow along with the fractures by water that cannot penetrate the deep formations because of their low permeability.

Turkey has important karst formations that cover wide areas. Economically, karstification and karst hydrogeology/hydrology are very important for Turkey. Six international symposiums have been arranged in the last 20 years in Turkey and have taken important steps in this region. UNDP has given significant support due to the importance of this subject, and DSİ (State Hydraulic Works) has supported these studies and projects together with Hacettepe University UKAM (International Karst Research Center). The scope of foreign engineers and hydrogeologists provided educational opportunities, and therefore, the world's largest dams in Turkey and groundwater projects performed.

Ankara, Turkey

Dr. Gültekin Günay
Emeritus Professor

Acknowledgments I would like to acknowledge and thank the following people for their great support and incentives during the preparation of this book. Prof. Dr. Petar Milanovic, Belgrad, Serbia; Dr. Jim La Moreaux, Pela, USA; Dr. Selim Yilmaz, Hacettepe University, Ankara, Turkey.

Contents

About the Authors

Dr. Gültekin Günay was born on January 2, 1939, in Karacailyas-Mersin, Turkey. He attended Mardin High School. He received his Bachelor of Science degree in Geology-Geophysics from the Faculty of Sciences of Istanbul University in 1962, followed by a Master of Science degree in 1968 at the same department. He later gained a Ph.D. degree in 1977 after completing his doctoral dissertation on the subject of "Isotope Hydrology" supervised by Prof. Dr. I. Enver Altınlı. He began his career in 1962 at Adana 6th District Directorate of the General Directorate of State Hydrological Works (DSİ). In 1966, he was granted a research scholarship by the International Atomic Energy Agency (IAEA) to continue his studies on Isotope Hydrology in the United States of America. In the United States, he conducted research under the guidance of the U.S. Geological Survey. Subsequently, in 1967, he continued his research in Vienna on Isotope Hydrology at the International Atomic Energy Agency (IAEA). From 1967 to 1968, he was employed as a hydrogeologist in the Department of Groundwaters, General Directorate of State Hydrological Works (DSİ). In 1976, he was appointed as National Project Coordinator of the DSİ-UNDP Project. During his tenure, he supervised the international karst hydrology investigations conducted comprehensively at a pilot project area of 16,000 km^2 in the Western Taurids. In 1978, he went to the United States as a "Visiting Scholar" and lectured at conferences in various American universities on the subject of "Karst in Turkey." He resigned from DSİ in October 1979 and began to work as a Faculty Member in the Department of Geology (Hydrogeology), Faculty of Engineering, Hacettepe University, Ankara.

His Associate Professorship Dissertation was on the Karst Hydrogeology of the Manavgat River Basin, and in 1982, he obtained his title of "Associate Professor." In the same year, he was appointed to the position of National Project Coordinator of the DSİ-UNDP project for the "Establishment of Hydrogeology Laboratories for Research and Education on Karst Water Resources."

He was appointed Full Faculty Member on November 30, 1984. Until 2000, he worked as the Chair of the Department of Hydrogeology at Hacettepe University. In 1985, he was appointed as the Director of the "International Research and Application Center for Karst Water Resources" (UKAM), and until 2000, he worked in this position with an outstanding performance. After receiving his "Full Professorship" in 1988, he was assigned as the Secretary-General of the Executive Committee of National Science Foundation (TUBITAK) Engineering Studies Group (MAG), a position which he occupied from 1989 to 1991. From 1991 to 1997, he was the Director of the Graduate School of Pure and Applied Sciences at Hacettepe University. In 2000, he retired at his own will from Hacettepe University and began to work freelance. Currently, he continues working as Hydrogeology and Geotechnical Advisor at Keystone Engineering.

He also worked as a Member of the Education and Instruction Committee of the International Hydrology Program (IHP) organized by UNESCO and served in the Turkish Delegation.

In 1985, he was selected for the membership of the International Association of Hydrogeologists (IAH) Karst Commission and also for that of the "International Association of Groundwater Engineers and Scientists."

He speaks English and Arabic. He is married and has two children.

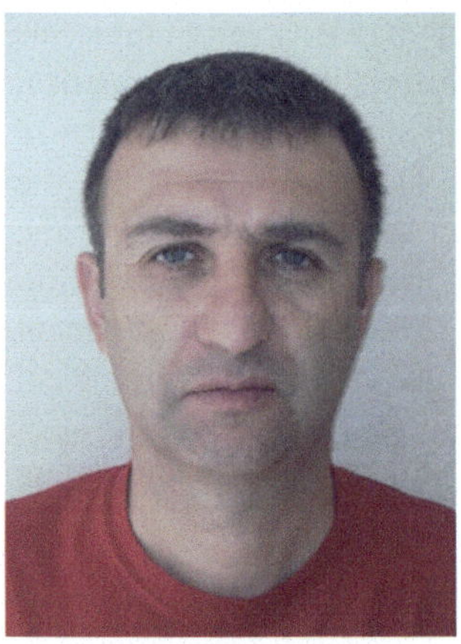

Koray Törk is originally a speleologist and karst hydrogeologist; he works most of the karst areas of Turkey. He was born in 1967 in Ankara. He joined to Cave Research Association (MAD) as a caver and then have been a member of Mineral Research and Exploration of Turkey (MTA) as a speleologist. He has been working predominantly on karst base natural hazards (sinkhole collapse/obruk) and cave protection for more than 10 years.

İsmail Noyan GÜNER was born in 1967 in Ankara. He graduated in B.Sc. as a Hydrogeology Engineer from Hacettepe University in 1993. He completed his M.Sc. and Ph.D. degrees in 1995 and 2009, respectively. He has been working since 1998 in MTA (General Directorate of Mineral Research and Exploration of Turkey) about hydrogeology of mining areas after working as a research assistant at Hydrogeology Department in Hacettepe University between 1995 and 1998. His professional skills are hydrochemistry, stable isotopes, and radiocarbon aging in groundwater, karst hydrogeology, and speleology.

Eric Gilli is a French karstologist. He was born in 1957 and studied geology and physical geography in Nice, Paris, and Aix-Marseille. Formerly a consultant in Nice, he has been a Professor in the Geography Department at the University of Paris 8 since 2001. His main topics are hydrogeology, karst, caves, groundwater, palaeo-environments, earthquakes, tsunamis, and natural disasters. He has been involved in shipwreck explorations and troglodyte studies. His international studies have taken him to many parts of the world.

Karst of Turkey

1

Gültekin Günay, Koray Törk, and İsmail Noyan GÜNER

Abstract

Approximately one-third of Turkey is covered with carbonate rocks. Intensive karstification in Turkey is seen in the whole karst areas. Intensive karstification heights of up to 2000 m are especially seen in the vicinity of Antalya city and generally in the Southwest of Turkey's Taurus Mountains. Especially, Antalya and its surroundings present such typical karstification properties. Kırkgözler karst springs in the north of Antalya plain are one of the few resources in the world with an average discharge rate of 15 m^3/s. Important progress has been achieved worldwide in karst hydrogeology studies in the last twenty years. In order to obtain a better understanding of the karst phenomena, classification of karst areas is a necessary and useful tool. Engineering geology studies are very important to understand and know the karst of the country.

1.1 Introduction

Alpine orogenic belt with the whole of Turkey is located in the Mediterranean region (Eroskay and Günay, 1980). Some important conclusions have been made in this region regarding Turkey's geology. The rock and sedimentary series that make up the country represent geosynclinal and orogenic facies and they are different from the peer series developed in the cratogenic shield areas and their shelf zones. Tectonic lines are orogenic origin. These structures are completely different from the formations that developed in the shield areas, which were subjected to no influence but slightly curved.

The very defective morphological surfaces, which are active in large parts of the country, are the result of intense tectonic movements and have not developed in the shield areas (Fig. 1.1).

Mesozoic has been affected by only one orogeny, namely the Alpine orogeny. Mesozoic in Turkey is composed of carbonate, radiolarite and clastics. The Cretaceous is one of the most important paroxysms of the Alpine orogeny, perhaps the most important of which is composed of carbonate and conglomerate and flysch series deposited in the Middle and Upper Cretaceous.

It is developed as "Gypsum Formation" in the Oligocene, Central, and North Anatolia. Miocene deposits were deposited in epeirogenic basins and in the Southeast Anatolia foreground. The large transgression from the Mediterranean Basin invaded South, South Eastern, and Eastern Anatolia (Kaçaroğlu et al. 1997).

1.2 Tectonic of Turkey

The building lines were carried out by this orogeny and the young epeirogenesis that followed it. In this case, some important facts emerge; All regions of Turkey have been under the influence of Alpine orogeny and young epirogenic movment. The structures, which were previously developed during the Caledonian and Hercynian orogenesis, were folded again, rejuvenated, or completely erased during Alpine movement. Only the very limited remains of these old buildings have been preserved between Alpine folds. However, not only the current tectonic lines, but also the relief (topography) of the country were brought into being by the Alpine orogeny and the epirogenesis that followed it (Bozkurt and Mittwede, 2001).

Figure 1.1 Simplified tectonic map of Turkey showing major neotectonic structures and neotectonic provinces

G. Günay (✉)
Faculty of Engineering, Hacettepe University, Ankara, Turkey
e-mail: ggunay@ttmail.com

K. Törk · İ.N. GÜNER
Hydrogeological Engineering, MTA Genel Md. Üniversiteler Mah., Dumlupınar Bulv. No: 139, Çankaya, Ankara, Turkey

G. Günay et al., *Caves and Karst of Turkey - Volume 2*, Cave and Karst Systems of the World,
https://doi.org/10.1007/978-3-030-95361-4_1

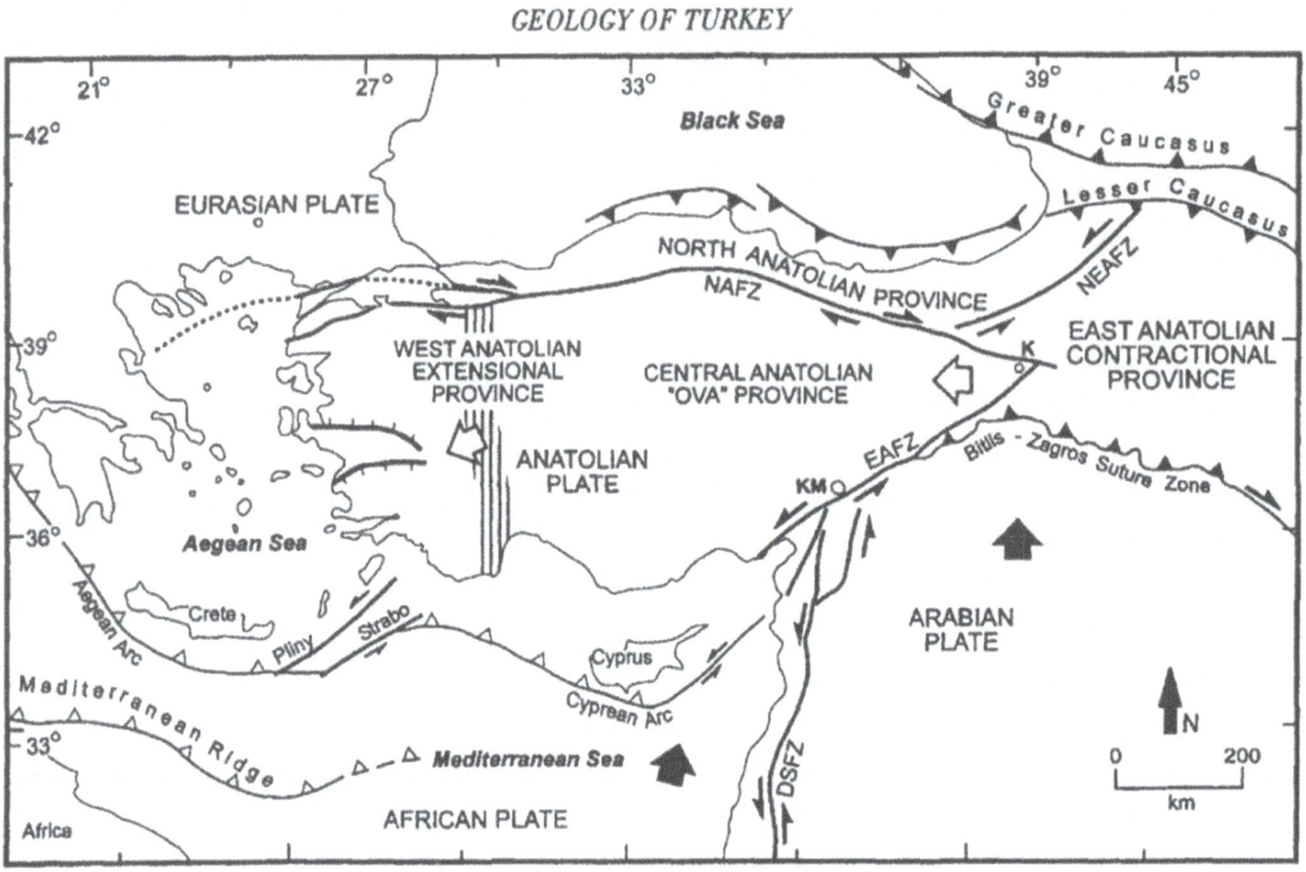

Fig. 1.1 International Geology Review, Vol. 43, 2001, p. 578–594. Copyright © 2001 by V. H. WINSTON & Son, Inc. All rights reserved (from Bozkurt 2001). Abbreviations: K = Karliova; KM = Kahramanmaras; DSFZ = Dead Sea fault zone; EAFZ = East Anatolian fault zone, NEAFZ = Northeast Anatolian fault zone. Heavy lines with half arrows are strike-slip faults. Half arrows show relative movement sense. Heavy lines with filled triangles show major fold and thrust belt; tips of the small triangles indicate an active subduction zone; tips of the small triangles indicate polarity. The heavy hachured lines show normal faults; hachure indicates downthrown side. Bold filled arrows indicate relative movement direction of African and Arabian plates; open arrows indicate relative motion of Anatolian plate. The hatched area shows area of transition zone between the Western Anatolian extensional province and the Central Anatolian "ova."

1.3 Tertiary Units

Tertiary desert deposits in some parts of the geosynclinal shale continued uninterruptedly into the Paleocene and then into the Eocene. Paleocene and Lower Eocene correspond to a rather quiet period that followed the Upper Cretaceous paroxysm. Compared to Upper Cretaceous, Paleocene-Lower Eocene formations have continuous and similar features. However, during the Cretaceous paroxysms, the topographic changes and exacerbated erosion that occurred in the inner part of the geosyncline also affected the accumulation conditions during the Paleocene-Lower Eocene. Therefore, sometimes very important facies differences are observed between the inner and outer parts of the geosynclinal area (Brinkman, 1976). Miocene deposits were deposited in epeirogenic basins and in the Southeast Anatolian foreground. The lithology and thickness of the series vary according to the size of the basins and tectonic developments. Besides the marine series left by the marine transgression, Lacustrine facies have developed (Atalay, 1987, 1988, 1996).

The Miocene had several important facies. Large transgressions from the Mediterranean Basin affected Southern, Southeastern, and Eastern Anatolia, where several large and many smaller Miocene basins were formed.

1.3.1 Karst in the Taurus Area

The Taurus Mountains are characterized by abundant water resources, large hydroelectric potential, some of the world's largest karst aquifers, and the largest karst springs. These springs of exceedingly large discharge are formed by the channelized flow along with the fractures by water that cannot penetrate the deeper formations because of their low permeability (Günay and Yayan, 1979; Karanjac and Günay, 1980). Many large springs occur at this contact and flow along the fractures by the water that cannot penetrate the deeper formations because of the low permeability. The original heterogeneity of the aquifer formed by tectonism has been increased by the dissolution of limestone. This dissolution results from the infiltration of a significant amount of rain and snow in the Taurus Mountains, particularly at higher elevations. If the transmissivity of the carbonate aquifers was more homogeneous, and the aquifers were hydrologically connected to the underlying sediments, the entire system's storativity would significantly increase, and diffuse flow would occur.

This area is characterized by abundant dolines, large poljes, coastal and submarine springs, sea caves, and large travertine terraces (Başar, 1972). One of the most unusual types of karsts is the one that is formed in an extensive conglomerate. The conglomerate is composed of cobbles of limestone with calcareous cement. The mode of transportation and deposition is not well understood (Erinç, 1960). As we can see at the Olukköprü springs area (Antalya—Beşkonak):

- The formation of karst in the Southwestern Turkey is strongly influenced by the alternation of the Jurassic and Cretaceous limestones with ophiolites and Tertiary flysch in both horizontal and vertical planes, resulting from nappe structures. Many large springs are where carbonate rocks are exposed within ophiolite.
- This region's general characteristics reflect a combination of both humid and arid environments, a combination that does not occur in a tropical climate. The unusual combination in Turkey results from the Mediterranean climate, which is characterized by dry summer, wet winters, and a long period of snow cover in the mountains. Therefore, mechanical weathering is an essential chemical corrosion.
- The hydrogeologic conditions of the karst indicate two types of recharge:
- Systems recharged by poljes on lower plateaus are characterized by a high fluctuation of discharge, higher temperature, higher bicarbonate content, and extensive travertine deposition;
- Systems recharged mainly from high mountains that are covered by snow for long periods are characterized by more stable regimes of discharge, lower temperatures, lower content of bicarbonate, and little or no travertine deposition. The Antalya travertine deposits were formed as a result of the precipitation of carbonate minerals due to the outgassing of carbon dioxide from the water initially saturated with respect to the carbonate minerals (Herman and Hubbard 1990). The source of the calcium and carbonate ions is the limestone of the Taurus Mountains to the north of the plain. The travertine in this plateau has a thickness of about 300 m and extends over an area of approximately 630 km^2. These deposits exist because of three separate terraces: one at an altitude of 300 m, another between 50 and 150 m, and the third terrace is below sea level.

A spectacular waterfall (the Düdenbaşı waterfall) on these terrace deposits has an unusual hydrogeologic occurrence (Back and Günay, 1992). The water travels to the waterfall through three different pathways even though the original source is one major group of springs. Among these springs, Kırkgözler springs are about 25 km from the waterfall. Even though a hydrologic connection exists between Kırkgöz springs and the Bıyıklı Sinkhole, the sinkhole receives water only during a high flood.

Part of the water from Bıyıklı and Yağca sinkholes travels to Varsak directly to the Düdenbaşı waterfall. It discharges into the doline where the water enters on the up-gradient side of the doline and also discharges down-gradient to continue to the Düdenbaşı waterfall.

The water source is from a canal that carries water from Kırkgöz springs to the Kepez power plant. After the water is used for electricity production, it is diverted into other canals and reused for irrigation. The returned irrigation water is transported by canal to the Düdenbaşı waterfall. The water cascades over a cliff about 15 m high, where an extensive cave system has developed. It is possible to walk in the open cave system beneath the waterfall. The Düdenbaşı River, formed by a waterfall, empties into the Mediterranean again with a waterfall. The Düdenbaşı waterfalls are a major tourist attraction in the Antalya area.

Another major tourist attraction in the hydrogeologically significant area is the spectacular travertine deposit of Pamukkale, which means "The Cotton Castle." It is on the ancient site of Hierapolis (an important city even before the time of the Romans). The travertine terraces are deposited from the series of geothermal springs brought to the surface along faults bordering the graben that forms the valley. The area has some potential for geothermal energy.

The major thermal spring source for the travertine terraces and ponds is a cave that was known as a road to Hades. The water from the thermal spring is 35 centigrade degrees, and originally, the spring was used for the medicinal qualities of the water. The atmosphere in the cave could not support life, probably due to the high concentration of carbon dioxide gas and consequent lack of oxygen.

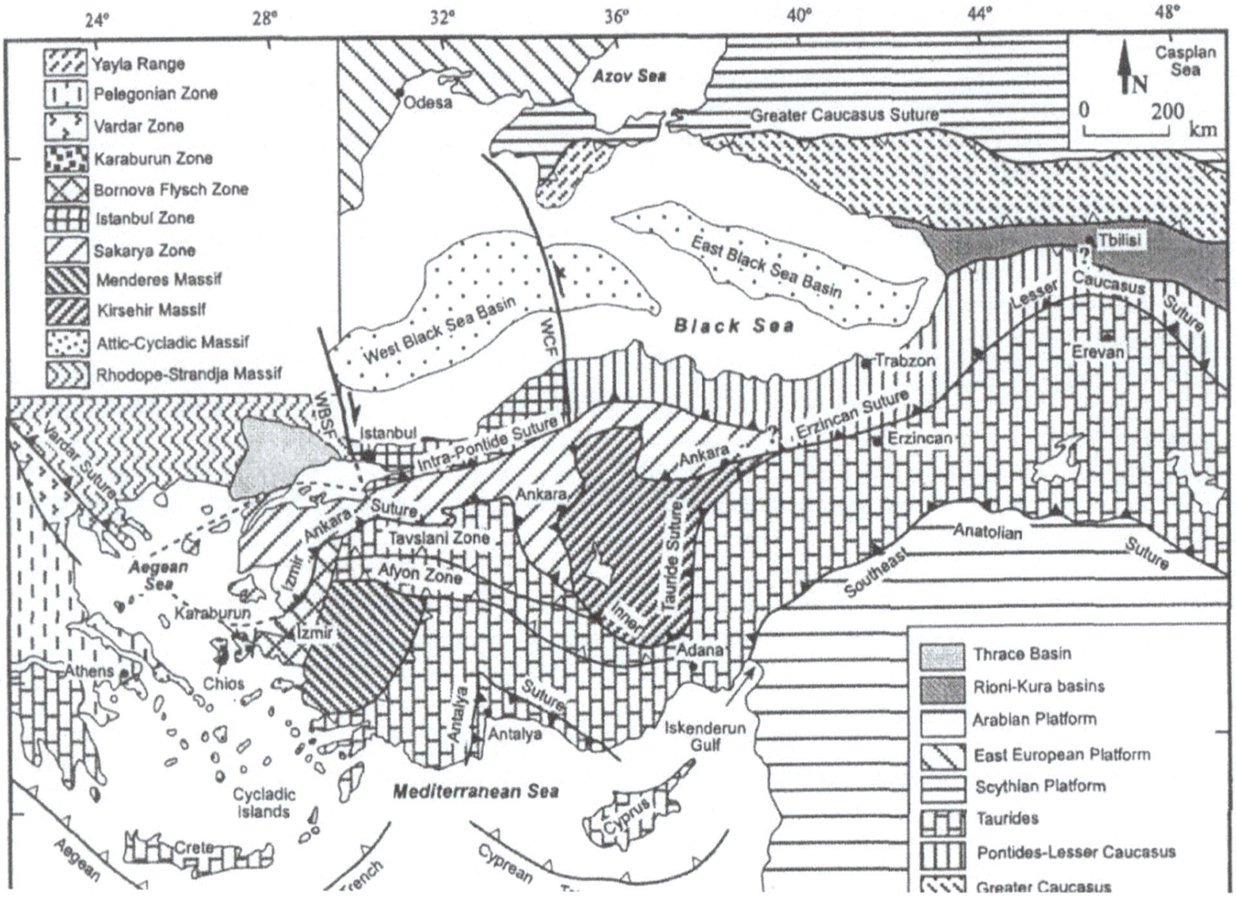

Fig. 1.2 Simplified tectonic map of the Taurus area. International Geology Review, Vol. 43, 2001, pp. 578–594.

The city's reputation was further enhanced by the miracles that the priests would perform by using the cave. It was well known that the cave was dangerous, yet the priests could stay in for a long period of time, presumably by following the pathways where the upper chambers contained adequate oxygen, and then emerge with no adverse reactions. That would then give them the aura of immortality, and they could then exert their will on people. In the second century AD, the Romans diverted the warm mineral water into the buildings for the baths.

The Roman bathhouse is now a museum and the thermal springs are available to tourists at hotels' swimming pools (Figs. 1.1 and 1.2).

Figure 1.2 Simplified tectonic map showing major neotectonic structures and neotectonic provinces modified from (Sengör et al. 1992; Barka 1992), abbreviations: wbs = western black sea fault; wcf = west Crimean fault. Heavy lines with filled triangles show sutures: the tips of triangles indicate polarity. Heavy lines with open triangles indicate thrust belts: triangles point toward the vergence direction. Heavy lines with half arrows show relative movement along these faults. The Pontides and lesser Caucasus form the eastern extension of the Sakarya zone (from Sengör et al. 1985)

References

Atalay, İ. (1987). Türkiye jeomorfolojisine giriş, 2. Basım. Ege Üniversitesi, Edebiyat Fakültesi Yayınları No. 9, İzmir (in Turkish).

Atalay, İ. (1988). Toros dağlarında karstlaşma ve karstik alanların ekolojisi. Jeomorfoloji Dergisi., 16: 1–8 (in Turkish).

Atalay, İ. (1996). Karstification and karstic landforms in Turkey, Karren Landforms, Eds. Fornos, Gines, Universitat de les illes Balears: 325–333.

Back, W., & Günay, G. (1992). Tectonic Influences on groundwater flow systems in karst of the southwest Taurus mountains, Turkey. in W. Back, J. S. Herman & H. Paloc (Eds.), Hydrogeology of Selected Karst Regions, International Contributions to Hydrogeology, Vol. 13, 263–272 p., Verlag Heinz Heise, Hannover, FRG.

Barka, A. A. (1992). The North Anatolian fault zone: Annales Tectonica, v. 6, p. 164–195.

Başar, M. (1972). Teşekkül tiplerine göre Türkiye mağaralarının Dağılışı, Jeomorfoloji Dergisi, Sayı. 1, 57–78 p., (in Turkish).

Bozkurt, E., & Mittwede, S. (2001). Introduction to the geology of Turkey - a synthesis, International Geology Review, 43, p. 578–594, Winston & Sons Inc., UK.

Brinkman, R. (1976). Geology of Turkey, Ferdinand Enke Verlag, Stuttgart, 158 pp.

Erinç, S. (1960). Konya bölümünde ve iç toros sıralarında karst şekilleri. Türk Coğrafya Dergisi, 20, 83–106 (in Turkish).

Eroskay, S. O., & Günay, G. (1980). Tecto-genetic classification and hydrogeological properties of the karst regions in Turkey. in G. Günay (Ed.), Karst Hydrogeology Proceedings: October 1979, Oymapınar-Antalya, Turkey, UNDP Project TUR/77/015, p. 1–41.

Günay, G. & Yayan, T. (1979). Antalya - Kırkgöz kaynakları hidrojeoloji incelemesi. 1. Ulusal Hidrojeoloji Semineri, DSİ Oymapınar Barajı, Antalya. DSİ-UNDP Projesi TUR / 77/ 015 Project Studies. DSİ Groundwater Dept. Yücetepe, Ankara (in Turkish).

Herman, J. S., & Hubbard, D. A. Jr. (1990). Travertine-marl: Stream deposits in Virginia, Department of Mines, Minerals and Energy Division of Mineral Resources, Vol. 101, 184 p., Charlottesville, VA.

Karanjac, J., & Günay, G. (1980). Dumanlı spring Turkey - the largest karstic spring in the world, Journal of Hydrology, 45, p. 219–231.

Kaçaroğlu, F., Değirmenci, M., & Cerit, O. (1997). Karstification in Miocene gypsum: An example from Sivas, Turkey. Environmental Geology, 30, (1/ 2), Springer-Verlag.

Şengör, A. M. C, Görür, N., & Şaroğlu, F. (1985). Strikeslip faulting and related basin formation in zones of tectonic escape: Turkey as a case study, in Biddle, K. T., and Christie-Blick, N., eds., Strike-slip faulting and basin formation: Society of Economic Paleontologists and Mineralogists, Special Publication, no. 37, p. 227–264.

Turkey's Karst and Water Resources

2

Gültekin Günay

Abstract

A considerable part of carbonate rocks is intensively karstified in Turkey. Water resources of the karst terrains of Turkey are rich and very important for the economy of the country. The gypsum karst units are also spread over a large area near the Sivas Province of Turkey, and especially in the east of Sivas. They are presumed to be from the Miocene age. The high mountain chains, often associated with Turkey's karst terranes, are responsible for these water resources' essential and beneficial characteristics. The strong orogenic movements and the Taurus Belt uplift caused the surface and underground water circulation. The complexity of the most karst regions of Turkey has led to the formation of large karst springs and aquifers.

2.1 Introduction

Turkey's karst limestone areas include very productive aquifers. The karstification in the limestones is common in Paleozoic, Cretaceous, Miocene, and Neogene formations. If the carbonate rocks at depths could be reached to withdraw their waters, Turkey's water resources development in the karst areas would likely reach up to 40% of the country's surface area. This simple percentage area covered by limestone, dolomite, and the other carbonate rocks and exposed to a potential solution of carbonates and karstification shows the importance of these rocks for Turkey's economic development. Most carbonate rocks of Turkey are intensively karstified.

G. Günay (✉)
Hacettepe University, Yaşamkent Mah. Ataşehir Sitesi 7A- 5. Blok No:12, Çankaya, Ankara, Turkey
e-mail: ggunay@ttmail.com

2.2 Gypsum Karst

2.2.1 General Overview in Gypsum Karst

The Sivas Basin of Central and Eastern Turkey is a classic area where the gypsum karst processes have been studied. Gypsum karst in Sivas Basin has developed in the Late Miocene at the Hafik formation, locally covered by Pliocene and Pleistocene sediments. The gypsum karst features are present in the region.

2.2.2 Gypsum Karst in Turkey

The gypsum karst units, which are spread over a large area in the vicinity of Sivas province of Turkey, and especially in the east of Sivas, are presumed to be of Miocene age. Geomorphological features include numerous collapse dolines.

The Seyfe and Göydün springs have high discharge rates of water unsuitable quality (EC: 13,000 ɥS/cm). Lakes are formed in the collapsed gypsum areas, and the water quality of these lakes is poor (Hafik lake, Tödürge lake, western Lota lake, eastern Lota Lake, etc.). The water quality of the Kızılırmak (Red) River, in this region, is not suitable for drinking and irrigation water.

The study area consists of sedimentary and metamorphic units of Paleozoic and Quaternary ages. In Figs. 2.1 and 2.2, the major Tertiary basin in Turkey can be seen. In Fig. 2.3, the generalized tectonic structures of the Anatolian plate is shown (Şengör, 1984).

Gypsum formations which contain rock salt (halite) interlayers crop out in a large area in Upper Kızılırmak River basin of the Sivas Province, Central Anatolia, Turkey. The Upper Kızılırmak basin covers an area of about 9,000 km2. The geological setting of the basin (Figs. 2.1, 2.2, and 2.4) is described mainly based on geological maps of MTA (General Directorate of Mineral Research and Exploration of

G. Günay et al., *Caves and Karst of Turkey - Volume 2*, Cave and Karst Systems of the World,
https://doi.org/10.1007/978-3-030-95361-4_2

Turkey). Metamorphic, magmatic and sedimentary rocks crop out in the study area whose ages range from Paleozoic to Quaternary.

Basement rocks consist of Paleozoic metamorphic rocks and magmatic rocks. Metamorphic rocks are mostly exposed in the North-Western part of the area, and consist of marble, recrystallized limestone, metatuff, metasandstones, Metavolcanics, quartzite, chlorite schist, sericite-schist, mica-schist, gneiss, and amphibolite. Magmatic rocks comprise granite, granodiorite, diorite gabbro, diabase, andesite, splite, porphyrite, basalt, and dolerite (Brinkman, 1976; Ketin, 1977).

Mesozoic rocks consist of Jurassic-Cretaceous massive, re-crystallized limestone; Cretaceous limestone, dolomitic limestone, sandstone, marl, marly limestone, conglomerate, and ophiolitic melange.

Paleocene-Eocene aged rocks comprise flysch (marl, sandstone, claystone, conglomerate), limestone, and volcanic rocks (andesite, basalt, agglomerate, tuff). Neogene rocks consist of Miocene sandstone, claystone, marl, clayey limestone, lacustrine limestone, gypsiferous series and volcanics (andesite, basalt, agglomerate, tuff); Upper Miocene-Pliocene sandstone, claystone, conglomerate, lacustrine limestone; and Pliocene conglomerate, sandstone, limestone and tuff intercalations. Miocene gypsiferous series comprise, conglomerate, sandstone, marl-gypsum alternation, massive gypsum, sandstone and claystone containing gypsum.

Quaternary deposits consist of alluvial, terrace deposits, and travertine. The alluvium is generally found along the valley and is composed of seeds of loose, interlayered clay, silt, sand, and gravel. Travertine crops out around some springs (especially hot springs).

Miocene gypsum, which contains rock salt (halite) interlayers, is the most extensive (about 50% of the area) lithological unit in the basin (Figs. 2.1 and 2.2) and is mostly karstified. Karst features developed in gypsum are dolines, ponors (swallow holes), solution pans (basins), depressions (troughs) of various sizes, and some caves. Some karstic springs issue from Miocene gypsum. Göydün (Ky-1) and Seyfe (Ky-2) springs constitute two main discharge points of the gypsum aquifer. The mean annual discharge of these springs is 1.10 m^3/s and 0.25 m^3/s, respectively (Kaçaroğlu et al., 1997).

The discharges of the Göydün and Seyfe springs do not change considerably between dry and wet seasons and are not directly affected by the precipitation's monthly variations.

The recession coefficients (α) of Göydün and Seyfe springs calculated by recession analysis were $2.55x10^{-4}$ and $1.65x10^{-3}$ day^{-1}, respectively. The quantitative data on the

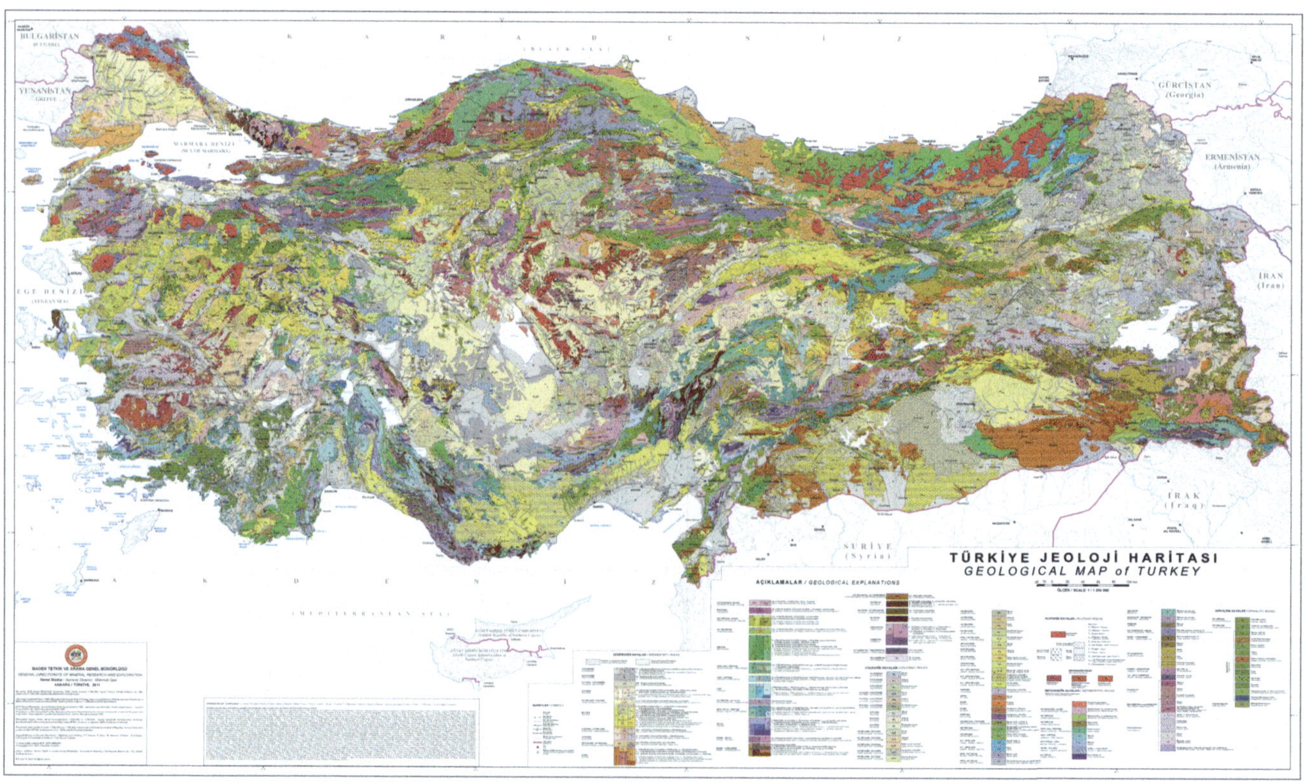

Fig. 2.1 MTA, 2011, Geological map of Turkey which is published in 2002 by Mineral Research and Exploration (MTA), reducing to the scale of 1/1.250.000 in a GIS environment, Ankara, Turkey

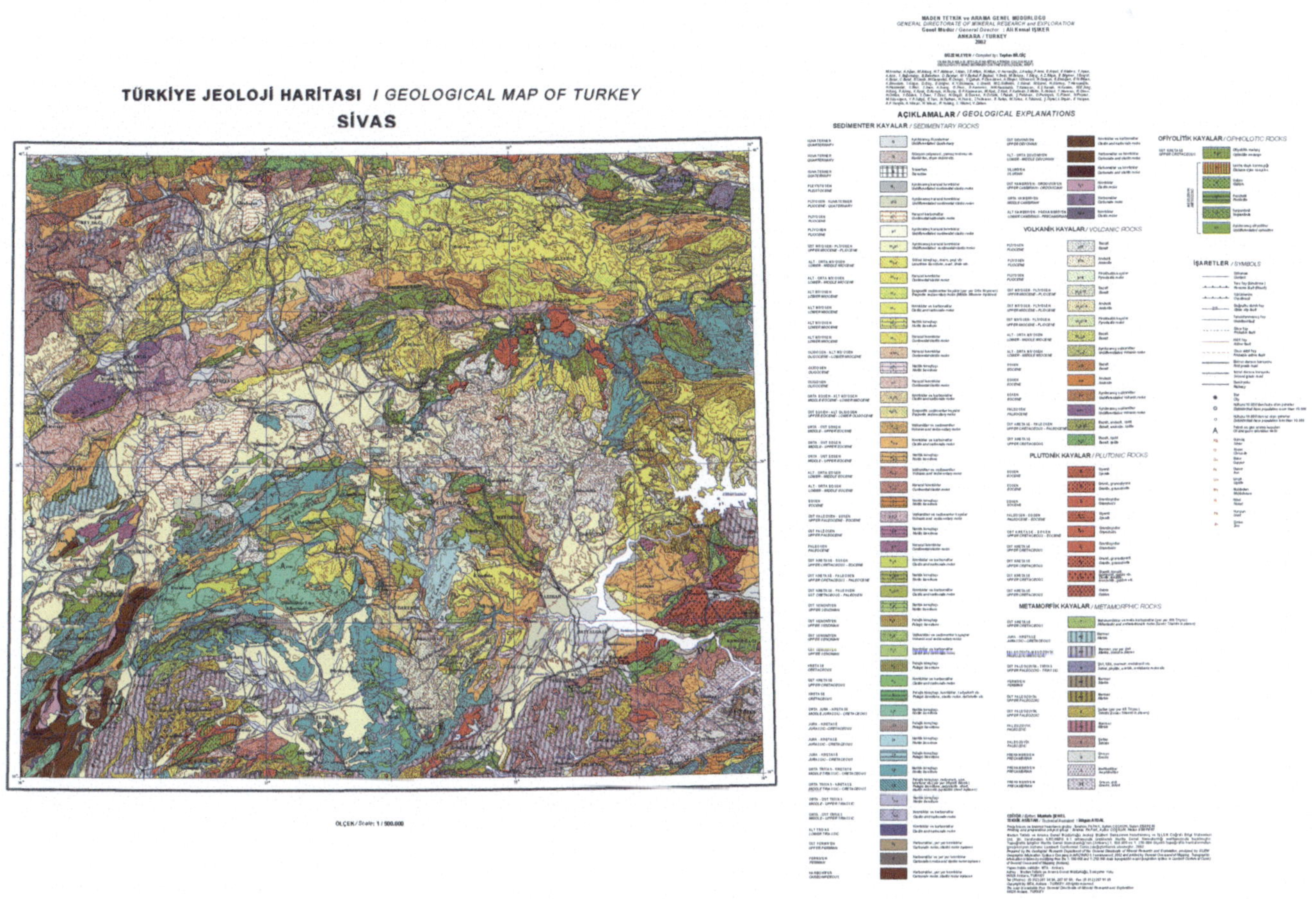

Fig. 2.2 MTA, 2002, Geological map of Turkey (1/500.000 scale) by Mineral Research and Exploration (MTA), in a GIS environment, Ankara, Turkey

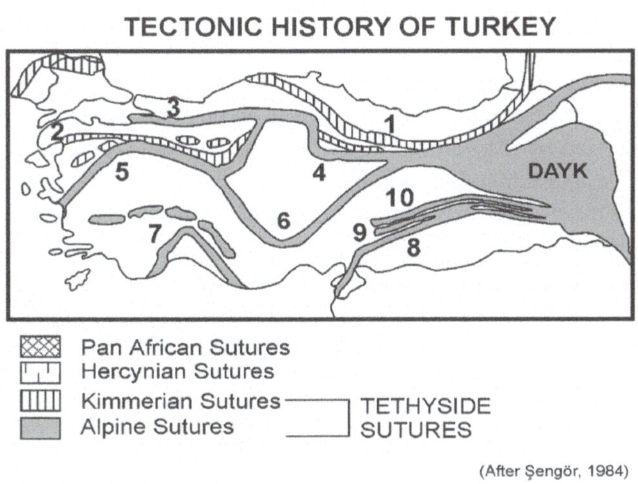

Fig. 2.3 Schematic map showing the distribution of suture belts in turkey. 1—Main paleo-tethyan suture, 2—karakaya suture, 3—intra-pontide suture, 4—erzincan suture, 5—izmir-ankara suture, 6—inner tauride suture, 7—antalya suture, 8—assyride suture, 9—çüngüş suture, 10—maden suture. Dayk: east anatolian accretionary complex. Suture, 5—izmir-ankara suture, 6—inner tauride suture, 7—antalya suture, 8—assyride suture, 9—çüngüş suture, 10—maden suture. Dayk: east anatolian accretionary complex

Fig. 2.4 General tectonic map of Turkey (MTA) (Emre et al. 2013)

groundwater velocities in the gypsum aquifer is not available. However, the recession coefficient and slight changes in the spring discharges indicate that the gypsum aquifer has a large storage capacity and groundwater flow occurs slowly, primarily through enlarged joints and fractures (Kaçaroğlu et al., 1997). Paleozoic marble and travertine in the northwestern part of the basin (in the Yıldız sub-basin) have aquifer characteristics and contain an essential amount of groundwater. The marble is thin to medium-bedded, partly massive, densely jointed, and joints are enlarged by dissolution.

The travertine is thin-bedded, vesicular textured, porous, and densely jointed. The water in these units is discharged by the large capacity Kaynarca (Ky-3) and Gaziköy (Ky-4) springs. Mean annual discharges of these springs are 0.60 and 0.23 m^3/s, respectively.

Neogene rocks (sandstone, conglomerate, limestone) and alluvium to the north of Sivas tavra valley are the main groundwater resources for Sivas (Figs. 2.1, and 2.2).

The gypsum karst units, which are spread over a large area in the vicinity of Sivas Province of Turkey, and especially in the east of Sivas, are presumed to be of Miocene age. Geomorphological features include numerous collapse dolines.

The Seyfe and Göydün springs have high discharge rates of water unsuitable quality (EC: 13,000 micro mho/cm). Lakes are formed in the collapsed gypsum areas, and the water quality of these lakes is poor (Hafik Lake, Tödürge Lake, Western Lota Lake, Eastern Lota Lake, etc.). The water quality of the Kızılırmak (Red) River, in this region, is not suitable for drinking and irrigation.

2.3 Karst Water Resources

The basic reasons for the large karstification of most carbonate rocks of Turkey are:

- The strong orogenic movements involving the carbonate rocks have lifted them much above the sea level and created significant level differences and powerful energy gradients for the surface and underground water circulation.
- Intense orogeny of folded, faulted, upthrusted, overthrusted and highly featured rocks provided both openings for the initial water circulation and opportunities for subsequent large rock solution and the creation of secondary porosity and;
- The Orographic processes and the resulting high mountain ranges represented barriers for the movement of air masses, forcing them to rise significantly and precipitate the rain and snow, and so provide large amount of water for the rapid infiltration, circulation, and solution of carbonates. (UNDP 1981).

Many external and internal factors are responsible for the type and the degree of karstification of an area of carbonate rocks (Daoxian and Back, 1991). However, the basic fact is that the geological structure, the orogeny and the connected tectonic provide the basic framework, which permits, enhances, or impedes the processes of karstification. The Alpine orogeny and the following epiorogenic movements in Turkey have been essential factors in karstification (Figs. 2.1 and 2.2).

The large karstification of carbonate rocks is spread almost all over Turkey. It is found mainly in the regions of the Taurus range, in western and south-western Anatolia, in Konya closed basin, in eastern Anatolia (Keban Reservoir region), and southern Anatolia (large Ceylanpınar plain) (Figs. 2.5, 2.6 and 2.7)Four different karst regions can be distinguished. Selected karst regions are: 1- Taurus Karst Region, 2- Southeastern Anatolia Karst Region, 3- Central Anatolia Karst Region, 4- Northwest Anatolia and Thrace Karst Region "Since the Taurus Karst Region is among the most important and largest karst area in Turkey, this article will deal briefly with the general geology of those karst regions (Eroskay and Günay, 1980; UNDP, 1981; UNDP, 1983).

2.3.1 Taurus Karst Region

The stratigraphy of the region contains various units from Cambrian to recent age. There are some para autochthonous metamorphites. carbonate units are mainly Devonian, Permian, Triassic, Jurassic, Cretaceous and Tertiary formations.Some units are local and cannot be assigned to the Taurus Mountains belt. The structure is relatively disordered. Necessary dislocations are mapped, with many upthrusts, overthrusts, carbonate and non-carbonate units reaching the thickness of thousands of meters (Şengör and Yılmaz, 1980). Among Paleozoic metamorphic rocks, there are recrystallized limestone, marble, and calcschist layers (Günay and Yayan, 1979). The schists; Triassic, Liassic, Cretaceous and Tertiary clastics, sandstones, shales, etc., generally form impervious and non-karstic barriers between the carbonate layers to the south generally cut longitudinal structural anomalies (Karanjac and Günay, 1980).

These zones of weakness provide connections between the carbonate belts. Mesozoic limestone and dolomite occur in over 1,000 m thick layers. Bentonic faces are widespread. The pelagic carbonates also occur. The evolution of the ophiolithic melange in the Late Cretaceous exists as the impervious base or cover according to the stratigraphical and structural position of carbonates units.

Important springs flow out of the mostly overthrusted limestone formations, located on impervious barriers. Non – carbonate rocks are also widespread, with impervious formations in the basement and ophiolitic melange, which often control the karstification development. Important karst features, which follow structural lineaments, are apparent. Karstic springs at the front face of over thrust are characteristic of this region. The karst of the Taurus region may be

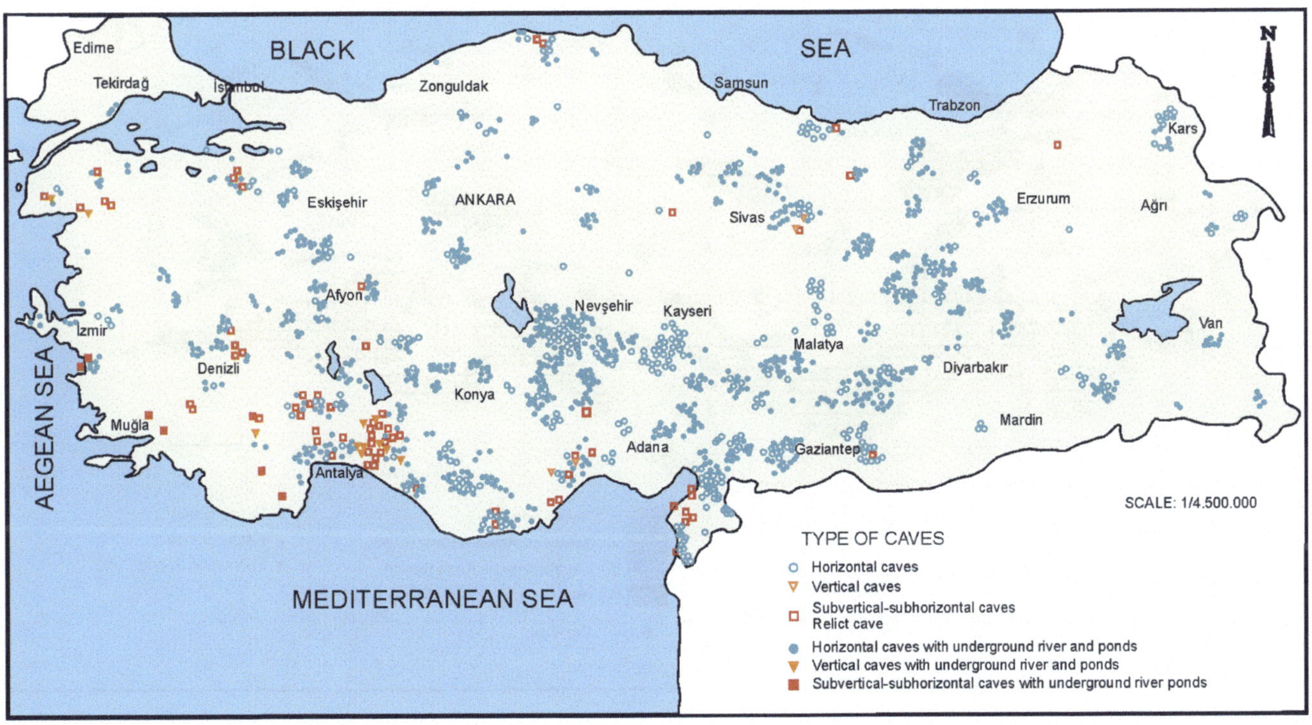

Fig. 2.5 Distribution of caves in Turkey (Başar 1972)

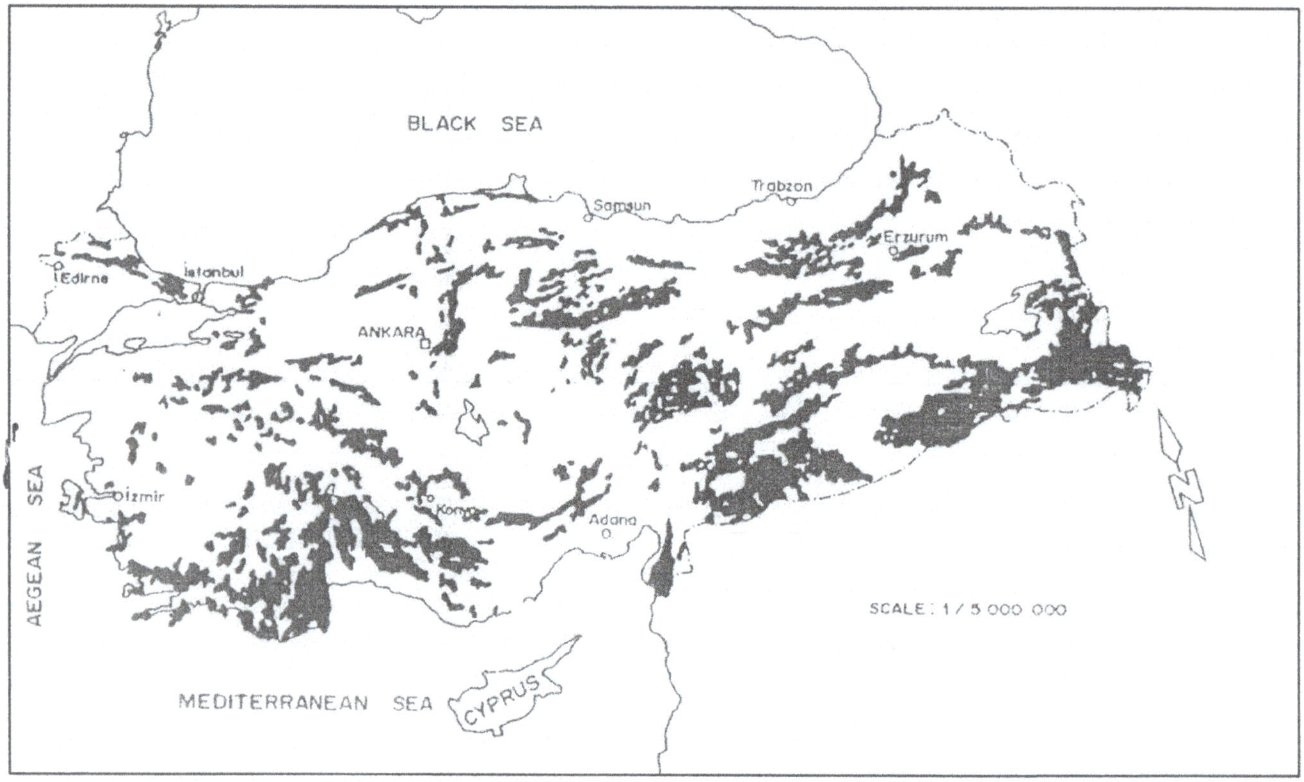

Fig. 2.6 Distribution of limestone formations in Turkey

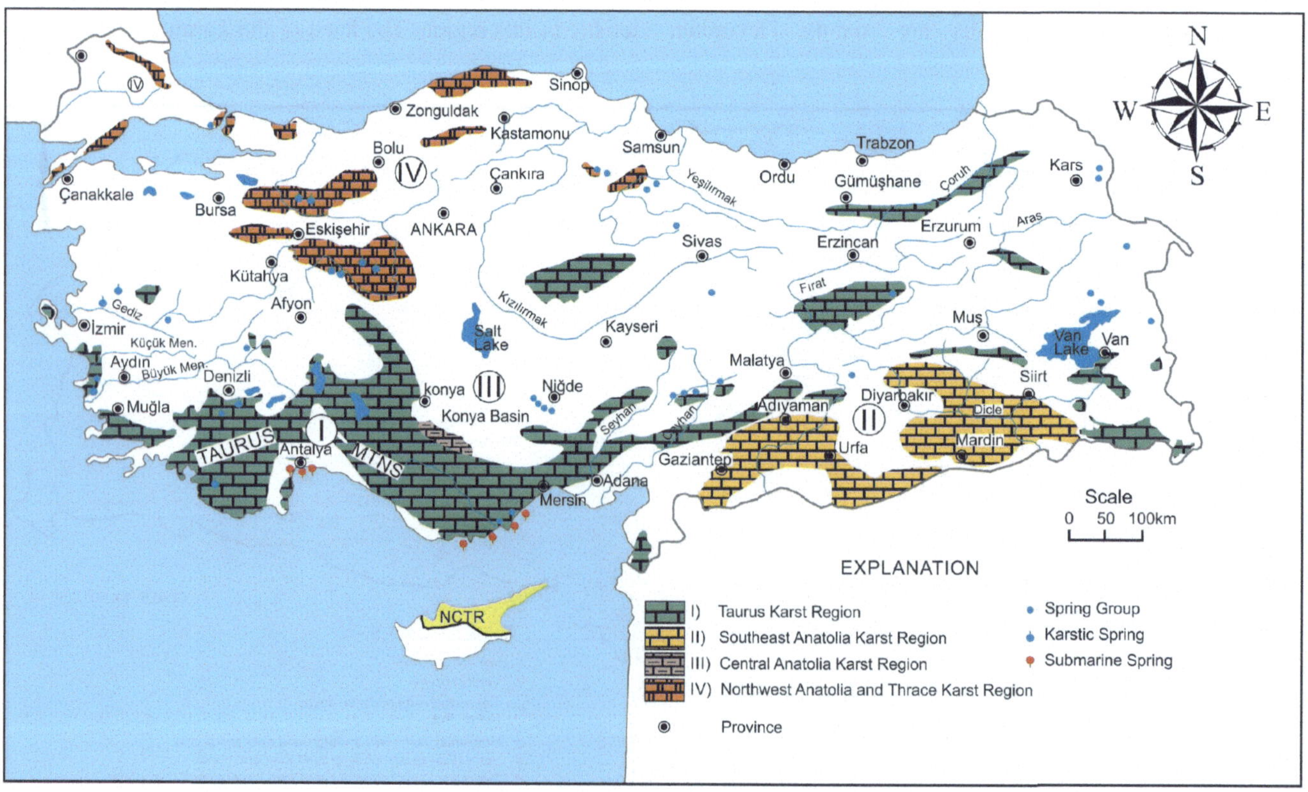

Fig. 2.7 Classification of Turkish karst (After Eroskay and Günay 1980—revised by Güner and Günay 2000)

attributed to "dissected orogenic karst." The western part of the Taurus Mountain may "correspond to orogenic accumulated karst." The middle part of the Taurus riders is highly influenced by the Paleozoic massif and the belt of Neogene impervious deposits, which prevent any contact of salt and fresh water. To the east, these belts are below sea level; this is why many coastal and submarine springs occur (Back and Günay, 1992).

The karst prototype of the Dinarides in Yugoslavia is of the same variety (Herak, 1977). The karst developed in the Miocene series between Silifke and Mersin corresponds to the "epi-orogenic tabular karst," while the karst developed in the Antalya travertines corresponds to "epi-orogenic deep karst," according to Herak classification. Carbonate plates of Libya may characterize "epi-orogenic tabular karst" and the karst of Jamaica "epi-orogenic-deep karst" according to Herak classification (Eroskay and Günay 1980).

2.3.2 Central Anatolia Karst Region

The Bozdağ limestone of the Permian age, which overlies metamorphic, covers a broader area. The recrystallized hard limestones of the Jurassic and Cretaceous age, which bound the South and West basin, are the extension of the Taurus. The ophiolite sediments of Late Cretaceous overly the older carbonate units. The Neogene units, which cover large areas in the basin, are mostly on an ophiolitic basement but in contact with older limestone. Neogene cover mainly consists of lacustrine clayey limestone or claystone. Neogene limestone is locally called "obruk limestone." The collapsed dolines in the basin, which are called "obruk", are typical karst features. There are obruk's which are developed by dissolution and collapsed and probably trace old dislocations such as Timraş, Kızören, Çıralı obruks etc.

Large karst springs issue mainly from Paleozoic and Mesozoic limestones, and some small springs issue from Neogene limestones. Some of these springs are: Çumra--Karapınar-Pınarbaşı springs 0.5 m^3/s; Karaman - Ayrancı - Akcaşehir springs 3 m3/s; Ereğli - Bor plain springs 6.5 m^3/s. The largest spring of the area is the İvriz Karst Springs, which issues from the Paleozoic karstic marbles with an average discharge of 5.7 m^3/s.

In the Upper Sakarya River basin, Sakaryabaşı springs issuing from Paleozoic limestones with the discharge of 3.6 m^3/s; Başkurt, Sadıroğlu and Göktepe springs issue from Neogene limestones with the discharge of 5 m^3/s (Günay, 2006).

A local and individual karst system exist here as Bozdağ limestone of the Permian age, which overlies metamorphics, covers a broader area. The recrystallized hard limestones of Jurassic and Cretaceous age, which bound the basin on the south and west, are the extension of Taurus, the ophiolite sediments of Late Cretaceous overly the older carbonate units. The Neogene units, which cover large areas in the basin, are mostly on an ophiolitic basement but in contact with older limestone. Neogene cover mainly consists of lacustrine clayey limestone or claystone. Neogene limestone is locally called Obruk limestone. The collapsed dolines in the basin, which are called "obruk" are typical karst features. Some Obruks are developed by dissolution and collapse and probably trace old dislocations such as is preserved at the lowest elevation of the basin.

In general, the region bounded with metamorphic units on the north and west. Among the schists, marble lenses of Permo - Triassic age (Günay, 1977). Neogene limestone is porous and thin-bedded. It is locally over Neogene limestone is porous and thin-bedded. Quaternary sediments locally overlay it.

Two different karst zones can be identified in this region. One of them is older formations of the Taurus belts, which occur at the margins of the basin, while the other one is gently dipping lacustrine Neogene karstic limestone, which is located in the middle of the basin.

The karst of the Central Anatolia (Konya closed basin and its surroundings) may correspond to the "Epiorogenic Basinal Karst." The bottom of the Konya basin consisting of "older karstified rocks" with the consequence of main or total subsurface drainage (Herak, 1977).

In the carbonates of young cover, widespread groundwater circulation is developed. The general hydraulic gradient in the basin for Neogene limestone is toward Salt Lake. Central Anatolia region morphologically seems to be a closed basin bounded with high mountains. The average elevation is around 1200 m. Salt Lake (Tuz Gölü) occupies the lowest elevation in this intra-mountain basin (Figs. 2.1 and 2.2). Two different karst zones can be identified in this region. The first is one of the older formations of the Taurus belt karst, which are seen at the southern margin of the basin, while the second one is Neogene lacustrine limestone karst, which is located in the middle of the basin. The karst of Central Anatolia, i.e., Konya closed basin and its surroundings may correspond to "epi-orogenic basinal karst". The bottom of the Konya basin consists of older karstified rocks with consequences of main or total subsurface drainage. In the carbonates of the younger cover wide groundwater circulation is developed. The general hydraulic gradient in the basin for Neogene.

2.3.3 South Anatolia Karst Region

The Gaziantep and Fırat clayey limestone formations show poor karstification. However, the same units with chalky appearance Cretaceous and/or older limestones outcrop and karstic features are widespread at the marginal fold belt. The surface drainage, which deepens from an approximately horizontal plateau, controls the subsurface karstification. For instance, near Dicle river, the old spring outlets can be traced all along with the river level from bottom to the karstic springs flow out at the Dicle river level, the water table is very deep generally, and diffuse groundwater flow is developed.

Vertical shafts that follow the joints are observed on sections of valley slopes. Some of them are filled with red soil. Large caves similar to those in the Taurus region are rare in the regional dip; groundwater flow is from north to south. Because of this, a very important discharge, the Rasal-Aïn (Syria) spring group can be seen near Turkey's southeast border. In this region, large karst springs issue mainly from the Eocene limestones. Only the largest ones will be mentioned. Rasal-Aïn springs on the Turkish–Syrian border 1 m^3/s on the Turkish side and 43 m^3/s on the Syrian side; Mardin - Hanik springs 0.5 m^3/s; Urfa - Harran springs 1.2 m^3/s; Upper Tigris (Yukarı Dicle) basin springs 10 m^3/s.

The southeast Anatolian karst with these peculiarities is distinguished from Taurus karst and represents another region.

The karst of the South Anatolia may correspond to the "Epi-orogenic Deep Karst" according to Herak classification.

2.3.4 North Anatolia and Thrace Karst Region

In this region, large karst springs are produced mainly from the Paleozoic limestones (marble), Eocene reef limestone, and Neogen limestones. In the Thrace parts of the region, the average discharge of the karst springs is 1.6 m^3/s; Kırşehir-Seyf springs is 0.3 m^3/s; in the Upper Sakarya River Basin, Sakaryabaşı springs issue from Paleozoic limestone with the discharge of 3.6 m^3/s; Başkurt, Sadiroğlu, and Göktepe springs issue from Neogene limestones with discharge of 5 m^3/s; Bursa and Çayırköy springs with 2 m^3/s; and Iznik—Orhangazi and Gemlik springs with 0.9 m^3/s.

2.4 Conclusions

In this region, individual lens-shaped limestone blocks covered limited areas when compared with other regions. As they show similar properties, Thrace and Northwest Anatolian karst region are considered together.

In the Thrace part, Eocene limestone of the southern part of the Istranca Massif is the important karst unit in the region. It extends as a thin belt nearly parallel to the massif. It is about 100–150 m. There is underline by impervious metamorphics and overlie clayey units of Tertiary age (Ketin, 1959). Reef limestones are indurated, hard, porous, thick-bedded, or massif, with solution cavities. Karst springs discharge near the south contact.

In this region large karst springs issue mainly from the Paleozoic Limestones (marbles), Eocene reef limestones and Neogene limestones, some of the smaller springs issue from Permo-Triassic, and Cretaceous limestones. The discharge rate of Bursa-Çayırköy springs and of İznik–Orhangazi spring are 2 m3/s and 0.9 m3/s respectively.

The limestone, parallel to the Istıranca Massif isocline fold, has an average dip of 20–30 degrees. On the impervious basement, the impervious basement units of the Central Sakarya area, permenant at the Kocaeli Peninsula, there are surface karst features on Triassic limestones that are 300–400 m thick. Mainly Triassic, Jurassic and Cretaceous limestones are located at different levels (Şengör et al., 1985).

Most of them are largely folded. Because of the structural positions indurated, hard and karstified limestones cannot form continuous karst systems. Some of them are hanging due to the regional uplifting and rapid erosion of river beds. Old outlets and natural bridges can be seen along Sakarya river benthonic limestones are more dissolved. Especially micritic, pelagic limestone levels do not show karstification. Generally, groundwater is stored in few places. In this region, karstification is strongly affected by structural factors.The karst of northwest Anatolia may be corresponded to the "Lenticular Orogenic Karst" type according to the "Herak Classification."

References

Back, W., & Günay, G. (1992). Tectonic Influences on groundwater flow systems in karst of the southwest Taurus mountains, Turkey. in W. Back, J. S. Herman & H. Paloc (Eds.), Hydrogeology of Selected Karst Regions, International Contributions to Hydrogeology, Vol. 13, 263–272 p., Verlag Heinz Heise, Hannover, FRG.

Başar, M. (1972). Teşekkül tiplerine göre Türkiye mağaralarının dağılışı, Jeomorfoloji Dergisi, Sayı: 4, Ankara, (in Turkish).

Brinkman, R. (1976). Geology of Turkey: Ferdinand Enke Verlag, Stuttgart, 158p.

Daoxian, Y., & Back, W. (1991). IGCP Project 299: Geology, climate, hydrology, and karst formation: Episodes, v. 14, no. 1, p. 80–81.

Emre, Ö., Duman.T. Y., Özalp, S., Elmacı, H., Olgun, Ş., & Şaroğlu, F. (2013). Türkiye diri fay haritası. MTA Genel Müdürlüğü, Özel Yayın Serisi-30. Ankara-Türkiye, (in Turkish).

Eroskay, S. O. & Günay, G. (1980). Tecto-genetic classification and hydrogeological properties of the karst regions in Turkey, in G. Günay (Ed.), Karst Hydrogeology Proceedings: October 1979, Oymapınar-Antalya, Turkey, UNDP Project TUR/77/015, p. 1–41.

Günay, G. (1977). Konya - Sarıcalar dolayının jeolojisi ve yeraltısuyu olanaklarının izotop yöntemlerinden de yararlanılarak incelenmesi, Doktora tezi, İst. Üniv. Tatbiki Jeoloji Kürsüsü, İstanbul, (in Turkish).

Günay, G. (2006). Hydrology and hydrogeology of Sakaryabaşı karstic springs, Çifteler, Turkey, Environmental Geology, 51, 229–240.

Günay, G., Güner, N., & Törk, K. (2015). Turkish karst Aquifers, Environmental Earth Science, 74, 217–226.

Günay, G. & Yayan, T. (1979). Antalya - Kırkgöz kaynakları hidrojeoloji incelemesi. 1.Ulusal Hidrojeoloji Semineri, DSİ Oymapınar Barajı, Antalya. DSİ-UNDP Projesi TUR / 77/ 015 Project Studies. DSİ Groundwater Dept. Yücetepe, Ankara (in Turkish).
Herak, M. (1977). Tecto-genetic approach to the classification of karst terrains, Krs Jugoslavije / Carsus Iugoslaviae, 9/4, 227–238, Zagreb, Yugoslavia.
Karanjac, J., & Günay, G. (1980). Dumanlı spring Turkey - the largest karstic spring in the world, Journal of Hydrology, 45, p. 219–231
Kaçaroğlu, F., Değirmenci, M., & Cerit, O. (1997). Karstification in Miocene gypsum: an example from Sivas, Turkey. Environmental Geology, 30 (1/ 2), 88–97 p., Springer-Verlag.
Kaçaroğlu, F., & Şahin, M. (1994). Tavra vadisinin (Sivas kuzeyi) hidrojeolojisi ve yeraltısuyu kalitesi, Ç.Ü. Yerbilimleri (Geosound) Dergisi, Sayı 24, s.117–133, (in Turkish).
Ketin, İ. (1959). The orogenic evolution of Turkey. MTA Dergisi, 3, 1–8 p., Ankara, Turkey.
Ketin, İ. (1977). Main orogenic events and paleogeographic evolution of Turkey, MTA Dergisi, 88, 1–10 pp., Ankara, Turkey.
Martinez, J., Johnson, K., & Neal, J. (1998). Sinkholes in evaporite rocks: surface subsidence can develop within a matter of days when highly soluble rocks dissolve because of either natural or human causes, American Scientist, 86(1): 38–51p.
Şengör, A. M. C. (1984). The cimmeride orogenic system and the tectonics of Eurasia, Geological Society of America Special Paper, 195, 82.
Şengör, A. M. C, Görür, N., & Şaroğlu, F. (1985). Strikeslip faulting and related basin formation in zones of tectonic escape: Turkey as a case study, in Biddle, K. T., and Christie-Blick, N., Eds., Strike-slip faulting and basin formation: Society of Economic Paleontologists and Mineralogists, Special Publication, no. 37, p. 227–264.
Şengör, A. M. C., & Yılmaz, Y. (1980). Tethyan evolution of Turkey - A plate tectonic approach, Tectonophysics, 75: 181–241.
UNDP. (1981). Karst waters of southern Turkey: final technical report of DSİ UNDP Project, in Turkey, TUR/77/015 Technical report, (prepared by V. Yevjevich), pp. 238.
UNDP. (1983). Strengthening DSİ groundwater investigative capability, phase II, Turkey, TUR/77/015 Project, DP/UN/TUR – 77–015/1 Technical Report, Karst Waters of Southern Turkey (prepared by V. Yevjevich).

Karst of Antalya Travertine, Southwest of Turkey

3

Gültekin Günay

Abstract

Antalya travertine plateau has an approximate area of 615 km^2. The average thickness of the travertine deposits is about 300 m. Several large springs exist discharging from the carbonates of the Taurus Mountain range of Southern Turkey. These large springs are characterized by great discharge rates (generally over 10 m^3/s) which have a large residence or turnover times of many years or decades of years, and large, well-regulated spring flows. The most important feature is the Kırkgözler karstic springs system, with an average yield being more than 15 m^3/s. The well-developed karstic features are deduced of the degree and the importance of the karstification in the travertine. Numerous sinkholes, springs, and depressions of various sizes are located in the travertine. There is a large submarine discharge of freshwater, and several coastal springs are known to exist.

3.1 Introduction

Most of the karstic aquifers of Turkey recharged from the high-altitude mountain regions. However, due to rapid groundwater circulation in conduits, the pollutants and contaminants could be conveyed quickly over large distances. Furthermore, the self-purification capacity of the karst aquifers is low compared to the granular aquifer systems (Daoxian and Back, 1991). Therefore, it becomes crucial to elaborate the knowledge for the karst system's functioning to protect these valuable resources.

Antalya travertine aquifer has been studied to determine the present state and the future trend of the pollution located on Turkey's Mediterranean coast. The aquifer's storage capacity is very high, and the aquifer is a potential water supply for the city of Antalya. The travertine is the deposition of a group of large karst springs called Kırkgözler located on the northwest edge of the study area. These springs discharge from the Mesozoic carbonate rocks with an average rate of 15 m^3/s. The thickness (−300 m) and the surface area (>600 km^2) of the travertine deposit are unusual when compared to similar formations in the world. The travertine has very well-developed significant primary porosity and karst morphology. Large solution channels, sinkholes, large karst springs, resurgences, and collapse dolines are the system's characteristic features.

3.2 Geographical Setting

Antalya travertine plateau is located at the Mediterranean coast of Turkey (Fig. 4.1) and bounded by the Aksu River Basin at the east, Beydağları Mountain at the northwest, and Antalya Bay at the south. The plateau has an approximate area of 615 km^2, and the average thickness of the travertine deposits is estimated to be 300 m. The plateau comprises three levels. The upper step of the plateau (Döşemealtı) starts on the foothills of Mesozoic carbonate rocks of Beydağları Mountains, which belong to the Western Taurus range. The Kırkgözler springs, recharging the travertine karst aquifer, emerge along a 1-km-long zone at the northwestern border of this step. The average elevation of the upper plateau is around 300 m. Antalya is located on the lower plateau called Varsak. The elevation of the lower step ranges between 40 and 150 m. There is an abrupt change in elevation of about 100 m between upper and lower levels. The third plateau is currently beneath the Mediterranean Sea, and its extension could not be precisely assessed.

The city's domestic water is provided mainly through karst springs and boreholes drilled in the travertine karst aquifer. The non-existence of a sewage system is one of the major problems that cause the degradation of groundwater quality. Domestic liquid wastes have been injected directly

G. Günay (✉)
Hacettepe University Eng. Faculty, Beytepe, Ankara, Turkey
e-mail: ggunay@ttmail.com

G. Günay et al., *Caves and Karst of Turkey - Volume 2*, Cave and Karst Systems of the World,
https://doi.org/10.1007/978-3-030-95361-4_3

into the travertine. This practice is expected to cause serious pollution problems in the near future. There are also various small settlements and villages over the travertine plateau using karst groundwater for venous purposes.

3.3 Geological Setting

3.3.1 Stratigraphy

In the Taurus range, lithological units are divided into two main groups: "autochthonous" and "allochthonous" formations, due to a very complex tectonic evolution. In the study area, autochthonous units comprise the Mesozoic carbonates and Cenozoic deposits (Ketin, 1966).

3.3.1.1 Autochthonous Units

Mesozoic

This unit is known as the "comprehensive series" and covers large areas. It comprises thick neritic carbonate rocks of Lias, Dogger, Malm, and Cretaceous ages. The bottom of these series crops out only in a limited area. Maastrichtian and Paleocene are represented by limestone formations, which are highly fractured and karstified.

Cenozoic

Pliocene deposits in the South Aksu Basin begin with clayey limestone and continue with continental conglomerate and sandstone. The total thickness is about 150 m. Quaternary is represented by Antalya travertine and alluvium. The travertine with an approximate thickness of 300 m covers an area of some 615 km^2 downstream of the Kırkgözler springs. Travertine depositions have been going on since the Plio-Quaternary period. Its porous and cavernous structure is mainly due to the sedimentary depositional conditions (UNDP, 1983).

The geological setting under the travertine deposits and the exact geometry of the contact between the travertine and the underlying Mesozoic carbonate rocks were not identified. This contact controls the groundwater flow from carbonate rocks into the travertine. During the drilling of an oil exploration borehole in Döşemealtı Area 7.5 km south of Kırkgözler springs called (Ismail -1) travertine (between 0 and 255 m), chert, limestone, shale, and radiolarite (between 255 and 340 m), and mafic plutonic rocks with fine texture (between 340 and 797 m) were identified.

3.3.1.2 Allochthonous Units

The allochthonous rocks belong to the Antalya nappes and are divided into three main units (Brinkmann, 1976). The impermeable nappe units constitute a significant barrier controlling the groundwater circulation.

Çataltepe Unit

This unit comprising the base of the Antalya nappes surrounds the eastern edge of the Beydagları massive. Its deposition has taken place between Upper Triassic and Upper Cretaceous. Upper Triassic is represented by clayey limestone and sandstone, whereas Jurassic and Cretaceous deposits are composed of thick neritic carbonates interbedded with radiolarite.

Alakirçay (Ispartaçay) Unit

The Alakırçay unit comprises siliceous limestone, radiolarite, submarine volcanic rocks intercalated with limestone and ophiolites of Upper Triassic and detritics of Upper Cretaceous (Günay, 1977). It overlies conformably Çataltepe unit. However, it is generally encountered over the limestone of the "Comprehensive Series" and forms an important impermeable barrier between the limestone and the permeable units of the upper nappe.

Tahtalidag Unit

Spreading over an area between the Eğirdir Lake and the Mediterranean Sea, it overlies the Alakırçay unit. The Tahtalidag unit, comprising carbonate and detritic rocks deposited between Cambrian and Cretaceous, reveals close similarity to the autochthonous limestone. The karstification is well developed in this unit and some large springs are discharging through it in the south of the study area. Alakırçay unit detaches this unit from the autochthonous carbonate rocks in the area.

3.3.2 Structural Geology

The Alpine orogeny phases predominating over the area have resulted in the development of significant folds (Şengör and Yılmaz, 1980). Nappes and numerous faults are brought by plate and fracture tectonics. The interpretation of the structural features helps to describe and understand the hydrogeological structure of the area.

Three different tectonic phases were identified in the Western Taurus range, namely Upper Cretaceous, Eocene, and Miocene tectonic phases. The Antalya nappes are overthrusted between the Upper Cretaceous and Eocene. The Eocene-Miocene deposits have later covered the nappes. Following these phases, post-Miocene tectonic movements were observed in the area, throughout the Pliocene and Quaternary deposits. Several lineaments and faults over the travertine were also seen from the Landsat images.

The previous studies in the area have concentrated on the emplacement of the Antalya nappes over the autochthonous unit. Three major lineament directions were detected in the study area. These are:

- Those extending in the NE-SW direction observed in the eastern side of the travertine and intersected the Upper Miocene and Pliocene deposits in the east.
- Those extending in the NW–SE direction, observed in the western site and intersected the Mesozoic limestone.
- Those extending in the N-S direction, intersected through the travertine, rare but continuously sited.

These directions generally coincide between the known groundwater flow directions and the distribution of the karst depressions in the travertine plateau. Major spring outlets are located on the intersections of the lineaments.

3.3.3 Paleogeography

The investigation area is situated within the Taurus tectonic range, a part of the Alpine orogenic belt. Formations from the Permian to present outcrop in the area.

The Lower Permian formations were thought to be deposited in shallow sea environments. Up to the Upper Permian, the sea was getting deeper, and dolomite of neritic facies was developed. At the end of the Permian, the area emerged for a short time. A transgressive limestone phase was followed by this period, and during the Triassic, the sea gradually deepened toward the Ladinian stage, when thick siliceous limestone and radiolarite were deposited (Şengör et al., 1985).

According to the paleontologic observations, the Liassic limestone was deposited in a relatively shallow sea extending to the continental shelf. In Dogger, many parts of the area emerged above the sea level, and in many places, Malm is observed as directly overlying the Permian, Triassic, and Liassic formations. During the Malm, the region was covered by calm waters for a long time, which enabled the deposition of thick layers of limestone and dolomite.

The Lower Cretaceous was characterized by shallow marine limestone again, and in the Upper Cretaceous, a deep sea covered the entire region once more until the Senonian. At the end of the Laramie phase of the Alpine orogeny, the sea regressed from this area, and no further evidence of marine facies was encountered (Ketin, 1977).

The Eocene was characterized by a warm, neritic, and epicontinental environment. In the Oligocene period, the climate was humid and warm, and very suitable for extensive karstification. After this period, the entire region was uplifted and allochthonous Antalya nappes were placed in the Early Miocene. During this emplacement, the Beydagları autochthonous carbonate unit was subsided to depths of about 500 m.

In the Pleistocene, coarse-grained calcareous conglomerates were deposited, and they were affected by strong tectonic movements. This phase was followed by peneplain development processes, which resulted in a new relief of the area, while spring waters were developed in the travertine (Back and Günay, 1992).

3.4 Karst Geomorphology

3.4.1 Travertine

The area is dominated by extensive travertine plateaus surrounded by steep mountains consisting of Mesozoic limestone with subordinate radiolarite and peridotite complex. The travertine occurs at two levels. The higher level lies at around 300 m above sea level and is known as the Döşemealtı plateau. To the east and south, a lower level, between 40 and 150 m above sea level, is known as the Varsak plateau.

The upper travertine plateau continues northward, into a plain underlain by radiolarite and isolated limestone outcrops. This plain terminates sharply at the edge of the limestone mountains.

The travertine plateau breaks out sharply with several escarpments toward the plain of the Aksu River eastward and to the coast southward. This results in some spectacular erosional features. Between the Aksu River and the travertine plateau, there is another plateau on Miocene conglomerate and claystone. The scarp between the Döşemealtı and Varsak plateaus is well pronounced and marked by sheer gorges more than 100 m deep. These gorges are supposed to have developed by dissolution and subsequent collapse of the overlying travertine roof.

The Karaman River system flows surficial across the Döşemealtı plateau over the impervious deposits of a large alluvial fan on the west side. The fan deposits terminate, and the river sinks into the collapse gorge, having a very steep thalweg gradient.

3.4.2 Karst Features

The degree and importance of karstification in the travertine deposits can be deduced from well-developed karstic features. Numerous sinkholes, springs, (Başar, 1972) and depressions of various sizes are located in the travertine. The most important feature is the Kırkgözler karstic springs system, with an average yield being more than 15 m^3/s.

The travertine plateau's origin and its lower members are ascribed solely to the precipitation of carbonates from these waters. The facies are essentially terrestrial, and marine fossils do not occur in the travertine. It is improbable that these plateaus have an origin as coastal terraces.

In the Kırkgözler area, a sizable natural lake has been formed by the emerging masses of water. Their surficial appearance is short-lived; water goes underground again through identifiable swallow holes to re-emerge after some 18 km as the Düden River at Düdenbaşı area. Before resurfacing, a karst window (Varsak) offers a view of the underground stream. The Düden River flows out as a surficial river over the Varsak plain, which forms a cliff of approximately 50 m at the coast. Much of the Düden River's water is used for irrigation, more in the past than at present, because of the growth of Antalya's urban area. A large portion of the Kırkgözler water is deflected over the Döşemealtı plateau to a hydroelectric power plant at Kepez, below the southern cliff. From there, a canal feeding an irrigation system contours the lower edge of the cliff toward Düdenbaşı, where the canal discharges into the resurging Düden River. On the Varsak plateau, irrigated agriculture is practiced wherever the layer of terra rosa and/or alluvial deposits is thick enough. At the foot of the Varsak coastal cliff, where it bends inland at Lara, a spring area called Kemeragzı marks the travertine's contact with the impervious Miocene claystone. It gives rise to a lake similar to that at Kırkgözler.

3.4.3 Poljes and Dolines

In the surrounding limestone of the Taurus Mountains, lineament analysis from remotely sensed images indicates a strong relationship with major and minor poljes, collapse sinks, and collapse gorges. The alignment of dolines is very striking in places; their lines often lead along lineaments to the borders of poljes (Nossin 1989).

3.4.4 Submarine Discharge

There is a large submarine discharge of freshwater, and several coastal springs are also known to exist. Three possible discharge points are supposed to exist well offshore, as indicated by a greater degree of turbidity in the seawater. The submarine springs and the submarine travertine plateau are associated with glacio-eustatic sea-level oscillations during the Pleistocene. A large "blue hole" is known in Antalya Bay. The existence of a submarine travertine plateau is also an indicator of relative sea-level movements.

3.5 Hydrogeological Characterization

3.5.1 General Hydrogeology

The main karst aquifers in the area are the Mesozoic carbonate unit and the travertine deposits. The allochthonous Antalya nappes control the boundaries of both aquifers to the west and the autochthonous Pliocene-aged clayey units to the east of the study area (Karanjac and Günay, 1980). The most important springs group in the area is Kırkgözler. This springs group discharges from the Mesozoic carbonate rocks along a zone of about 1 km length through limestone and travertine contact.

Being the major karstic unit in the area, the Mesozoic carbonate unit with a thickness of over 1000 m exhibits all characteristics of the Mediterranean type of karstification (Herak,1977). Due to intensive karstification, the secondary porosity is well developed. The lithological and structural features generally control the solution cavities and the conduits. The karstification in Mesozoic carbonates should have begun long before the travertine depositions. The spring water discharging from the Mesozoic carbonates is highly supersaturated with respect to calcite. Due to the physical and biogenic degassing of CO_2 at the outlet of the springs, travertine has been precipitated (Herman and Hubart, 1990; Herman and Hubart, 1992).

The travertine aquifer that has been formed by the carbonate precipitation from the Kırkgözler springs is also recharged by the same springs. Some of the surface outflow of Kırkgözler springs diverted for hydroelectric power production and irrigation. The remaining part goes into the travertine through identifiable swallow holes at Bıyıklı. In addition to this amount, based on some speleologic and tectonic investigations, the travertine is believed to be fed from the Mesozoic carbonates beneath the surface.

Various paleo-springs and caves located in the Mesozoic carbonate unit, above the orifices of the Kırkgözler springs, are accepted as apparent evidence for an ongoing karstification process. Before the development of presently discharging springs at around 300 m elevation, the Mesozoic carbonates should have been drained through these paleo-springs. The ongoing karstification process extending down into deeper parts of the carbonate unit is controlled by the present sea level and allows a recharge from the Mesozoic rocks into the travertine aquifer (Denizman, 1989).

The recharge of the travertine aquifer is mainly controlled by the recharge area of the Kırkgözler springs. Therefore, the catchment area is not limited to the travertine plateau and extends to the catchment area of Kırkgözler springs, which is estimated as 4300 km^2 (Tezcan 1993). The rainfall over the travertine plateau is another recharge source for the aquifer.

The mean annual rainfall is 1.060 mm over the travertine plateau, and due to the highly porous structure of the travertine, much of this amount infiltrates rapidly underground.

The discharge of the travertine occurs through several springs at different levels.

3.5.2 Kirkgözler Springs

The Kırkgözler springs emerge 300 m above the sea level through numerous outlets, along a zone of 1 km long at the southern foothills of Katran Mountain, 30 km north to Antalya. Their recharge area comprises the Beydagları karstic limestone. These springs emerging through the fractures of limestone along the zone mentioned above create a swampy zone in front of the discharge area. The average discharge rate of the springs is about 15 m^3/s (Figs. 4.2, 4.3 and 4.4).

The karstification developed parallel to tectonic movements, and the impermeable Çataltepe unit acting as a barrier for the groundwater flow has played an important role in the development of Kırkgözler springs. In the proximity of springs, despite the relatively thin travertine, which is in contact directly with limestone, the detritic rocks, alluvial material, and the highly porous travertine allow the groundwater to flow from the limestone aquifer into the travertine. The part of groundwater exceeding the underground solution conduits' flow capacity emerges at the surface to form Kırkgözler springs (Figs. 3.1, 3.2, 3.3 and 3.4).

3.5.3 Düdenbaşi Spring

Düdenbaşı spring discharges through the travertine in the east of the lower plateau with an average rate of about 17 m^3/s. The origin of this spring is a matter of debate. The dye test earned out in 1977 has proved the connection between the Bıyıklı swallow hole and the Düdenbaşı spring through Varsak doline. However, the discharge rate of the Düdenbaşı spring is generally higher than that of the Kırkgözler springs. The spring responds very rapidly to rainfall events due to the highly porous structure of the travertine. In rainy seasons, discharge rates up to 150 m^3/s have been recorded. The recession of these peaks is also very rapid. One of the major questions is the occurrence of this spring and the origin of its water. The water of this spring flows through the Düden stream to the sea (Fig. 3.5).

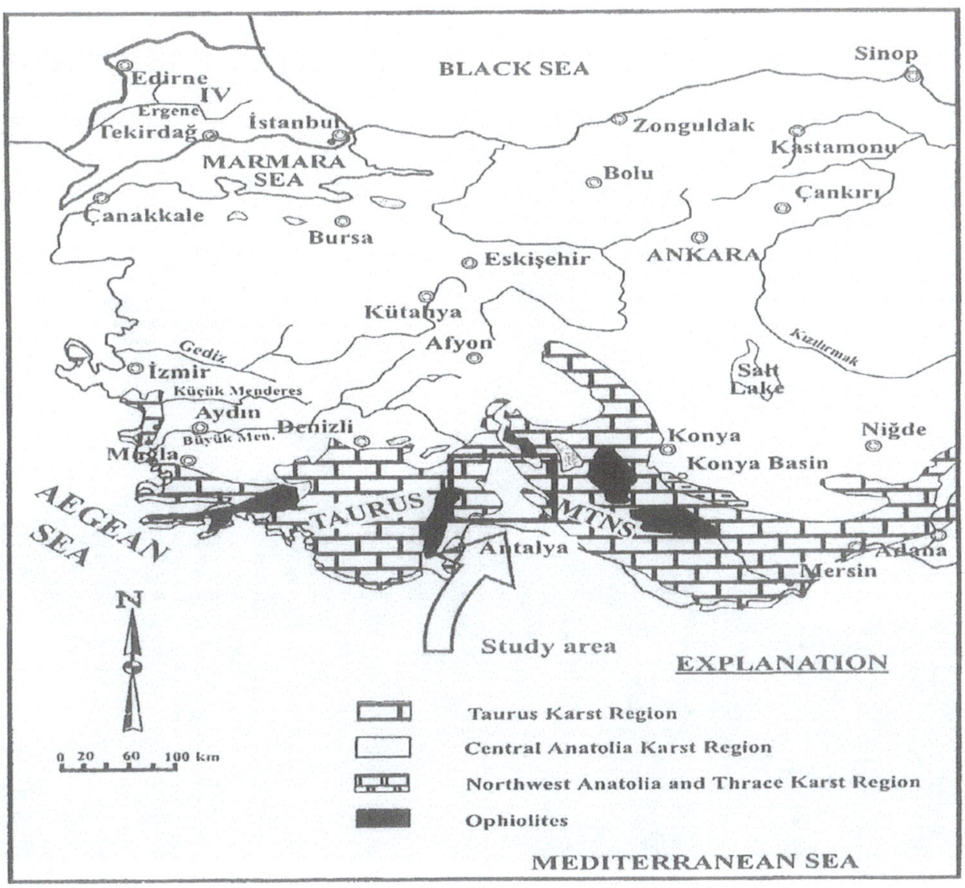

Fig. 3.1 Distribution of the karstic rocks and location of study area in Turkey (Eroskay and Günay, 1980)

Fig. 3.2 Kirkgözler springs 1

Fig. 3.3 Kirkgözler springs 2

Fig. 3.4 Kirkgözler Springs

Other springs are concentrated on the southwest of the lower plateau (Günay and Yayan, 1979). Their discharge rates, compared to the Kırkgözler and Düdenbaşı springs, are relatively low. The occurrence of these springs and their relation with the Kırkgözler springs could not be identified yet. The discharge rates of these springs have declined in recent years due to the groundwater's extensive use through the pumping wells in this area. The Magara spring located in the west of Antalya has been utilized for water supply to the city with an average rate of 1.5 m^3/s. The outlet of this spring is 2 ^{m}/asl, and no seawater intrusion problem has been encountered. Numerous coastal springs exist at the seashore. In 1979, 59 outlets were located by State Hydraulic Works (DSİ), and their total discharge rate was estimated to be 5 m^3/s (DSİ 1985).

Fig. 3.5 Antalya Düdenbaşi waterfalls and Düdenbaşi underground river

References

Back, W., & Günay, G. (1992). Tectonic Influences on groundwater flow systems in karst of the southwest Taurus mountains, Turkey. in W. Back, J. S. Herman & H. Paloc (Eds.), Hydrogeology of Selected Karst Regions, International Contributions to Hydrogeology, Vol. 13, 263–272 p., Verlag Heinz Heise, Hannover, FRG.

Başar, M. (1972). Teşekkül tiplerine gore Türkiye mağaralarının dağılışı, Jeomorfoloji Dergisi, Sayı 4, Ankara (in Turkish).

Brinkmann, R. (1976). Geology of Turkey: Ferdinand Enke Verlag, Stuttgart, 158p.

Daoxian, Y., & Back, W. (1991). IGCP Project 299: Geology, climate, hydrology, and karst formation: Episodes, v. 14, no. 1, p. 80–81.

Denizman, C. (1989). Kırkgöz kaynaklan ve Antalya traverten platosunun hidrojeolojik etüdü, Yüksek mühendislik tezi, Hacettepe Üniversitesi Fen Bilimleri Enstitüsü, Beytepe, Ankara, 159 sf. (in Turkish).

DSİ. (1985). Antalya Kırkgöz kaynakları ve traverten platosu karst hidrojeolojisi etüt raporu, DSİ Genel Müdürlüğü, Yücetepe, Ankara (in Turkish).

Eroskay, S. O., & Günay, G. (1980). Tecto-genetic classification and hydrogeological properties of the karst regions in Turkey. in G. Günay, (Ed.), Karst Hydrogeology Proceedings: October 1979, Oymapınar-Antalya, Turkey, UNDP Project TUR/77/015, p. 1–41.

Günay, G. (1977). Konya-Sarıcalar dolayının jeolojisi ve yeraltısuyu olanaklarının izotop yöntemlerinden de yararlanılarak incelenmesi, Doktora tezi, İ.U.F.F. Tatbiki Jeoloji Kürsüsü, İstanbul (in Turkish).

Karanjac, J., & Günay, G. (1980). Dumanlı spring Turkey - the largest karstic spring in the world, Journal of Hydrology, 45, p. 219- 231

Günay, G. & Yayan, T. (1979). Antalya - Kırkgöz kaynakları hidrojeoloji incelemesi. 1. Ulusal Hidrojeoloji Semineri, DSİ Oymapınar Barajı, Antalya. DSİ-UNDP Projesi TUR / 77/ 015 Project Studies. DSİ Groundwater Dept. Yücetepe, Ankara (in Turkish).

Herak, M. (1977). Tecto-genetic approach to the classification of karst terrains, Krs Jugoslavije / Carsus Iugoslaviae, 9/4, 227–238, Zagreb, Yugoslavia.

Herman, J. S., & Hubbard, D. A. Jr. (1990). Travertine-marl: Stream deposits in Virginia, Department of Mines, Minerals and Energy Division of Mineral Resources, Vol. 101, 184 p., Charlottesville, VA.

Herman, J. S., & Hubbard, D. A. (1992). The role of ground water in the deposition of travertine-marl, in W. Back, J. S. Herman and H. Paloc (Eds.), Hydrogeology of Selected Karst Regions, International Contributions to Hydrogeology, Vol. 13, 263–272 p., Verlag Heinz Heise, Hannover, FRG.

Ketin, İ. (1966). Tectonic units of Anatolia (Asia minor), MTA Dergisi, 66(1), 23–34 pp., Ankara.

Ketin, İ. (1977). Main orogenic events and paleogeographic evolution of Turkey, MTA Dergisi, 88, 1–10 pp., Ankara, Turkey.

Nossin, J. J. (1989). SPOT stereo interpretation in karst terrain, Southern Turkey, ITC Journal, 2, 79-91.

Şengör, A. M. C., Görür, N., & Şaroğlu, F. (1985). Strikeslip faulting and related basin formation in zones of tectonic escape: Turkey as a case study, in Biddle, K. T., and Christie-Blick, N., eds., Strike-slip faulting and basin formation: Society of Economic Paleontologists and Mineralogists, Special Publication, no. 37, p. 227–264.

Şengör, A. M. C., & Yılmaz, Y. (1980). Tethyan Evolution of Turkey: A Plate Tectonic Approach, Tectonophysics, 75: 181-241.

Tezcan, L. (1993). Karst akifer sistmlerinin trityum izotopu yardımıyla matematisel modellemesi, Doktora tezi, Hacettepe Üniversitesi Fen Bilimleri Enstitüsü, Beytepe, Ankara, 125 sf. (in Turkish).

UNDP. (1983). Strengthening DSİ groundwater investigative capability, phase II, Turkey, TUR/77/015 Project, DP/UN/TUR – 77–015/1 Technical Report, Karst Waters of Southern Turkey (prepared by V. Yevjevich).

Konya-Karapinar Sinkholes (Obruks) of Turkey

4

Gültekin Günay and Ilhami Çörekçioğlu

Abstract

The karstic features in the Konya-Karapınar plain located in the Central Anatolia region of Turkey are very interesting and very important. The number of watery and dry sinkholes (obruks) formed in the lacustrine Neogene limestone in the Karapınar plain is close to thousands. Groundwater with unconfined aquifer characteristics appears close to the surface and form sinkhole lakes by taking part in some of the sinkholes. The basement formation underlying the Neogene limestone is Mesozoic crystalline limestone and hydraulically related to the Neogene limestone. While the groundwater level in the sinkholes here is close to the surface, in recent years, the unplanned groundwater discharge has dropped to 90–100 m and many watery and dry sinkholes have started to form in the plain. Neogene limestone has a porous structure and contains solution cavities. One of the main factors in the formation of sinkholes in this plain is the presence of limestone with pores and solution cavities, the presence of clayey, sand materials, and solution cavities formed by the carbon dioxide (carbonic acid) groundwater formed by the effect of nearby Hasandağı old volcano.

4.1 Introduction

The term "obruk" (sinkhole) is a special Turkish term given to the cylindrical dry and watery sinkhole formed in Konya-Karapınar plain in Central Turkey (Ford and Williams, 1989). Such structures were later seen in the Urfa region in southeastern area.

Some of these sinkholes are large scale and are generally watery sinkers. Some of them are dry sinkholes of smaller diameter and not too deep watered sinkholes are of great importance for Konya-Karapınar plain. Because the plain is a region with very little rainfall and there is a need for irrigated agriculture in the first plan, the irrigation process, which was carried out by drawing groundwater from watery sinkholes, reduced the groundwater level in the Karapınar plain in a short time and as a result, sinkholes started to form in the plain (Güldalı and Şaroğlu, 1983).

The work started with the warning of the public in time, but when the low education level could not speak to the public, the water level in the sinkholes also decreased and many drilling wells were dried. At this stage, studies have been initiated to transfer water to the Konya-Karapınar plain through water channels to be opened from other basins, for example the Beyşehir Lake Basin (Günay, 1980).

4.2 Geology

4.2.1 Konya-Karapinar Plain

The mountains around the Konya plain consist mainly of crystallized limestones. Above these layers are flat-bottomed conglomerate and Bozdağ limestone with light to dark gray, partially recrystallized and visible cracks. The limestone formation was dated back to the Permian age with the help of few fossils. The Bozdağ limestone overlaps a thick sequence of light-colored, compact, solid, and Jurassic-Cretaceous aged limestone, which can also be seen along

Published in: Carbonates Evaporates (2010).

G. Günay (✉)
Faculty of Engineering, Hacettepe University, Beytepe, Ankara, Turkey
e-mail: ggunay@ttmail.com

I. Çörekçioğlu
Hydrogeologist DSİ Regional Directorate, Konya, Turkey

G. Günay et al., *Caves and Karst of Turkey - Volume 2*, Cave and Karst Systems of the World,
https://doi.org/10.1007/978-3-030-95361-4_4

with the Taurus Mountain range in the south (Fig. 5.1). The plain also features an extensively exposed Cretaceous ophiolite suite and thick Tertiary formations. In the east, there are different sedimentary basins of the Paleocene, Eocene, and Oligocene ages, as well as lacustrine deposits of the Miocene and Pliocene ages, which cover large areas in the plain. The overall thickness of the plain with the above-mentioned layers is about 350 m, with local variations (70 m at Çıralı and 150 m at Kızören Sinkhole). The thickness of the plain is even less near to Salt Lake (Tuz Gölü) (Geologic map Fig. 4.1, 4.2).

4.2.2 Geology of Obruk Plain

The sedimentary environment during the Mesozoic age was a large and deep oceanic basin. Because of the v Upper Cretaceous uplift, depositional characteristics were changed, and all of the units were folded. This movement is attributed to the first phase of Alpine orogeny that caused a regional uplift through which the Bozdağ Paleozoic sequence was exposed to surface erosion (Ketin, 1977). Following the regional uplift, deposition in Konya Basin has changed from flysch to lacustrine continental character during the Upper Eocene and Lower Oligocene lacustrine sediments included some evaporitic deposition during the Oligocene (Şengör, 1980). The Miocene Basin is represented mainly by limestone deposits. This limestone is covered uncomfortably by lacustrine detritus of Pliocene age. This discontinuity represents the Attic Phase of the Alpine Orogeny (Günay, 1977).

The widespread lacustrine Neogene deposits comprise conglomerates, marlstones, siltstones, claystone, and limestone. According to the drillings, data obtained from the Turkish Petroleum LTD., claimed that the thickness of the limestone in this formation reaches (Canik and Çörekçioğlu 1985) up to 300 m in some places.

4.2.3 Structural Geology

There are peculiar relationships between the groundwater conditions and the fold and fault patterns in the area. The fold patterns are not noticeable due to the recent sedimentary sheet. However, the hydrogeological trends have been explicated and are shown on the hydrogeological map in the NW–SE direction.

4.3 Hydrogeological Units

This research was carried out to understand the local groundwater potential, and differentiation of hydrogeological units was preferred for this purpose. The rationale for this approach is that it allows a better understanding of the differentiation of permeability limits, groundwater motions, hydrological conditions, and the nature of sinkholes. The product of the combination of these parameters is a simple hydrogeological model of the region (Günay et al., 2010).

The impermeable basement complex to the north, northeast, and west surrounds the Konya plain. Karstic limestone and dolomite of the Permian and Jurassic-Cretaceous ages, varying between 300 and 1200 m thick, form a uniform aquifer together. The impermeable barriers seen in the region consist of ophiolites. However, as can be seen from the water wells' lithological logs, it is seen that they are not in a continuous layer from their surface.

The folding arrangement of the bedding is not very intense. The volcanism in the region cuts or enters the sedimentary strings. As can be understood from geophysical studies, Tertiary and Quaternary, formed alternately by permeable and impermeable deposits, form a layer with a total thickness of 10–15 m.

Claystone, clayey or silty shale, and sandstone, alternating toward the southeast and Ereğli depression, form Paleocene, Eocene, and Oligo reed sequences. Large-scale faulting affected these exposures.

The aquifers in the region are formed by Pliocene-aged lacustrine limestone forming a cover affected by solution channels. Intermediate layers with clay can also be seen in the area. The obruks are in these non-uniform limestones. The volcanic in the region contains large impervious units.

In this study, precipitation, replenishment, and the relationship between them were studied in the old obruks. The water levels of some sinkholes such as Timraş, Kizören, and Meyil were measured twice a day. Precipitation data were obtained from the local weather station's records. Data from Konya, Aksaray, Karaman, and Karapınar stations were also taken into account to understand the regional precipitation trends better.

The parallel between the May and September rainy season curves indicates a correlation between the Çarşamba stream discharge and the precipitation. The sudden decrease in the curvature in the June-July period is important because the water tables in the sinkholes are not parallel to these trends. The reason for this situation is the continuous feeding from irrigation channels. Irrigation channels are located on horizontally permeable (KA2) limestones, and leakage from these channels contributes to the local water level and thus explains the relationship between the water level of the obruk and the local water table. However, it is not reasonable to believe that these fluctuations can induce the entire groundwater body since the sinkholes are located at a depth of 30–40 m or even 150 m from the local topographic surface. Finally, it should be noted that regional hydraulic gradients and boundaries control the horizontal retention of groundwater.

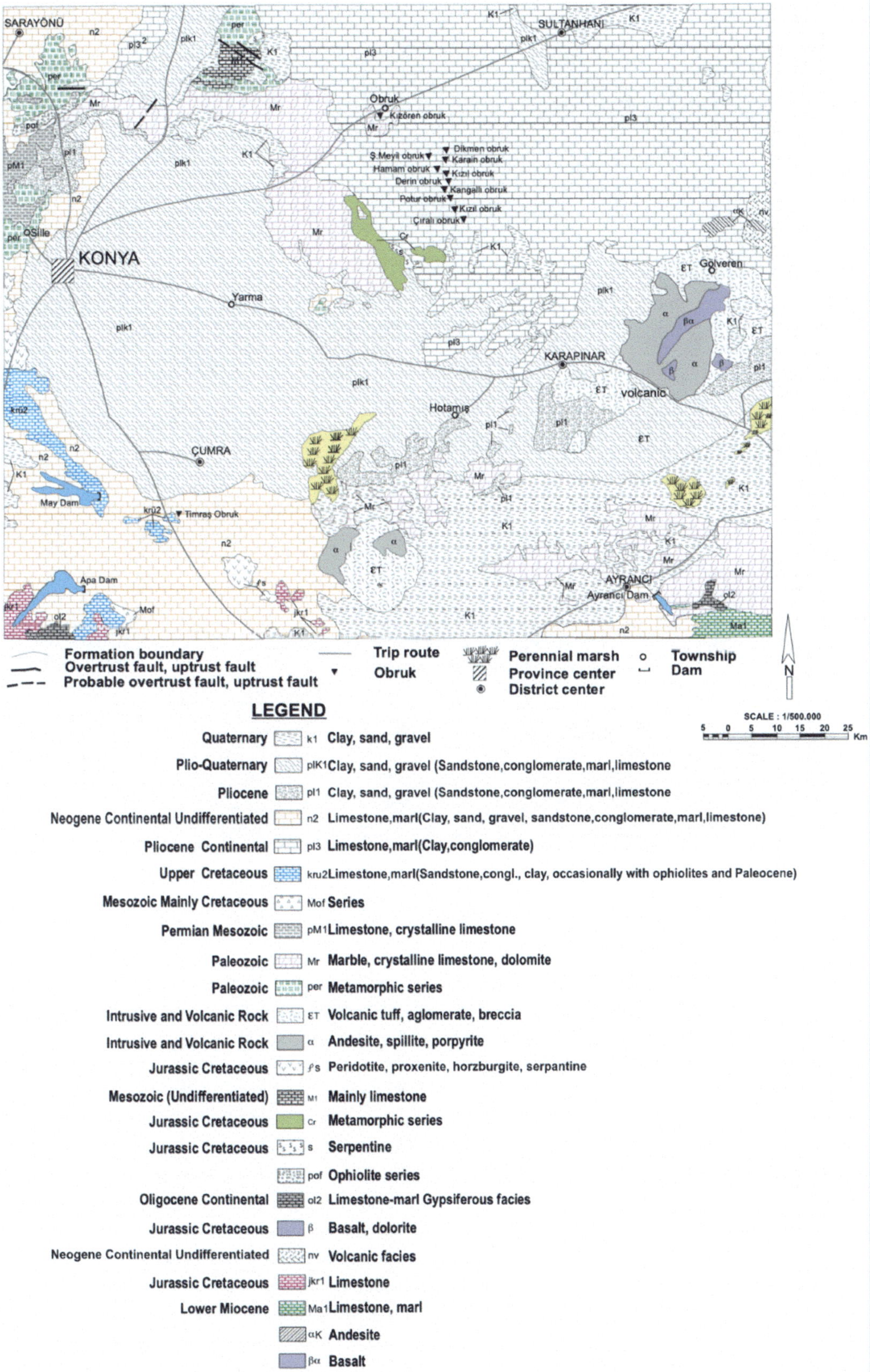

Fig. 4.1 Simplified geological map of the Karapinar Obruk plain area (after Çörekçioğlu 1994)

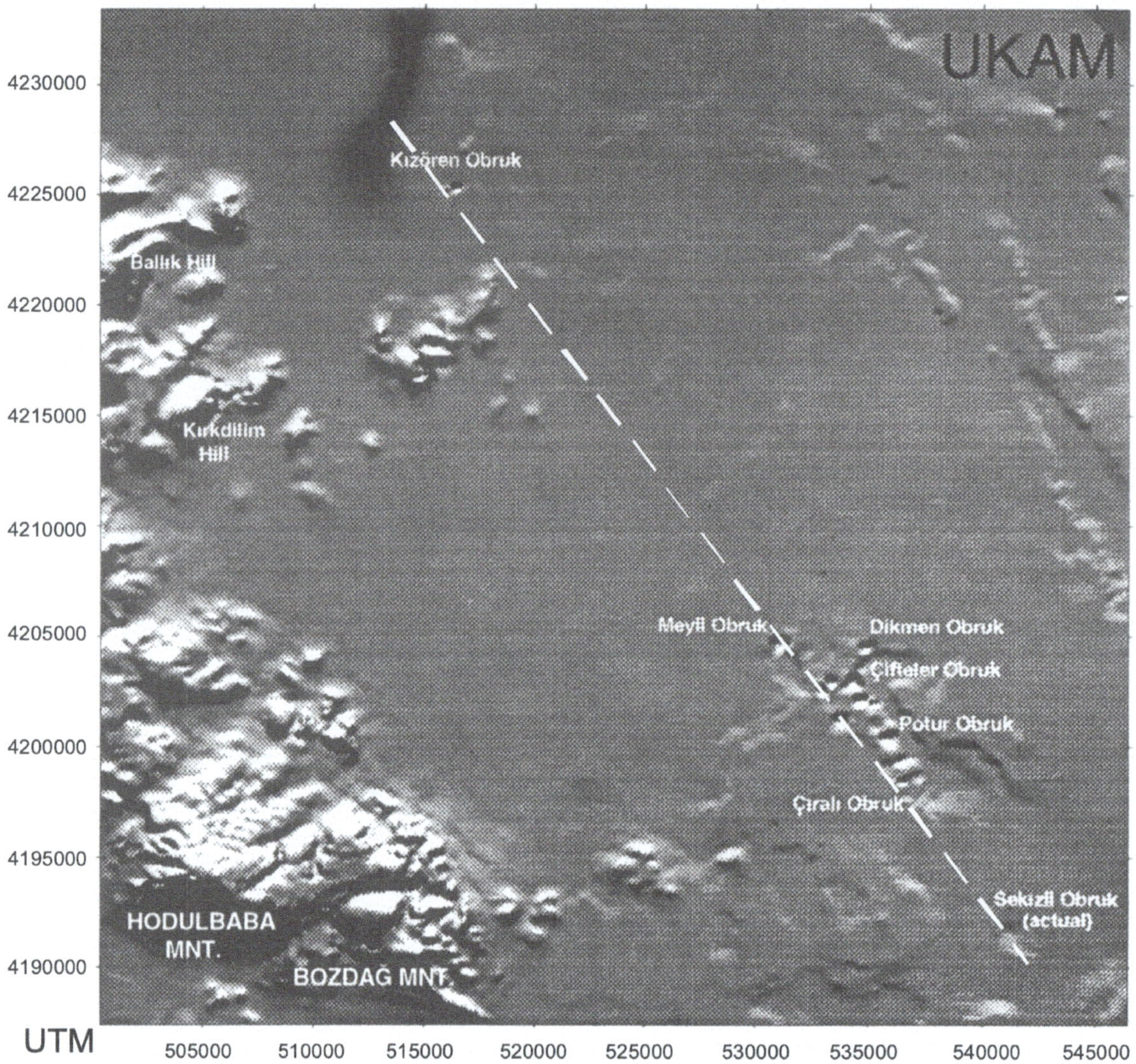

Fig. 4.2 Illumination image of digital elevation model of Obruk Plateau of Konya closed basin (Çörekçioğlu 1994; Günay 1995)

Geophysical studies have also confirmed some significant displacements that form an impermeable barrier in the eastern part of the plain. The water levels of Kizören and Meyil sinkholes put pressure on them (Hydrology Figs. 4.3, 4.4 and 4.5).

4.4 Obruk's Development

4.4.1 Origin of Obruks

Superficially, sinkholes are found in Neogene limestones. The vertical cylindrical-elliptical shape of the obruks can be explained by assuming that the basement floor of the caves coincides with the upper parts of the crystallized limestones. However, this assumption still needs to be confirmed. The lower and upper sectors are probably located within different but uniform carbonate rocks. This type and mismatch can also be observed in terms of transferability in the horizontal and vertical directions.

The obruks appear to conform to certain trends and concentrate on fault lines. The last situation can be seen in Kızören, Meyil, and Çıralı sinkholes. Although the sinkholes (Gökhöyük, Apasaraycık, Belkuyu, etc.) located in the southwestern part of the map are in NW-SE direction, they do not coincide with the visible faults. They also extend along the Neogene limestones strike, where the origins of their formation are only speculative. Pressurized water can be considered as an alternative explanation. However, this alternative explanation requires an assumption about the structural subsidence of cap limestone and the higher permeability for lower limestone—still the overall permeability value. Moreover, solution channels in Neogene limestones are largely clogged with debris that cannot be easily dissolved.

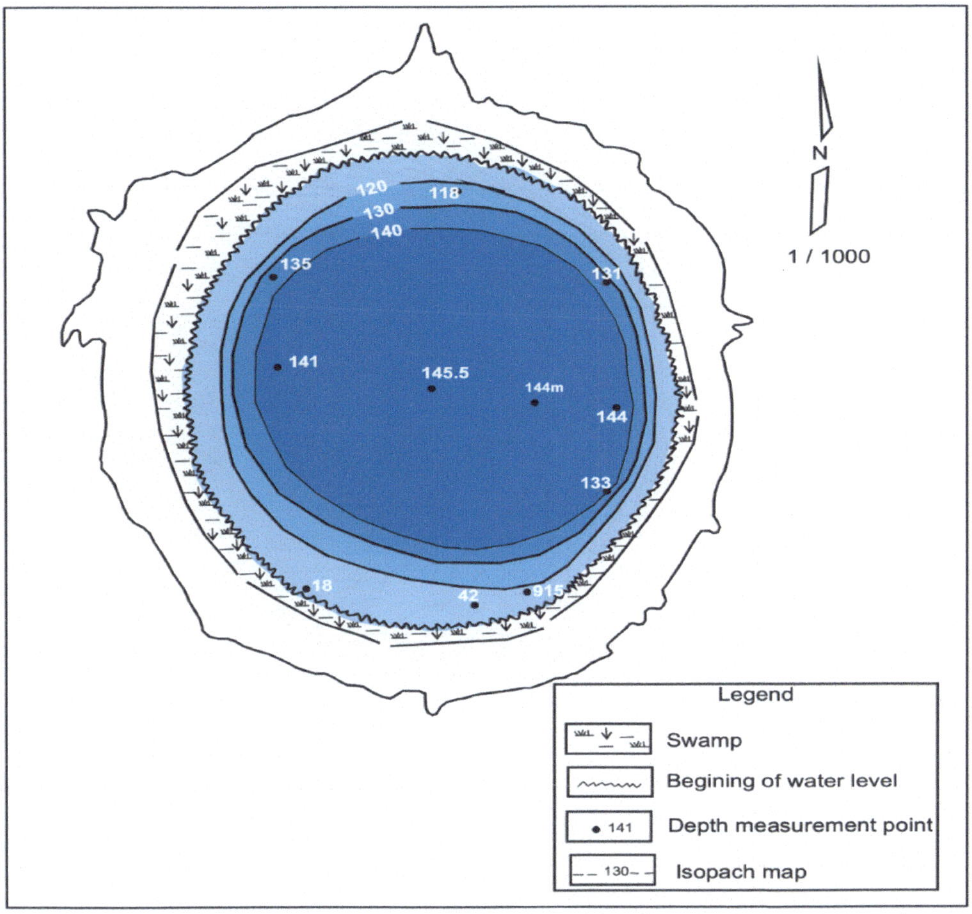

Fig. 4.3 Batthymetric map of the Meyil Obruk (after Çörekçioğlu 1994)

It is possible to assume that the country rose from the water level from the alluvial environmental fans. Under this assumption, the movement in Neogene limestones can be explained as a rearrangement.

It is understood from the analysis of water samples taken from various depths that there is a gradual enrichment in H2S. The reason for this enrichment may be a slower movement or recharging over an extended period of time. The direction in which waters flow is defined by local variations that occur during ascending motion (Eroskay, 1976).

The distribution of the obruks on the surface does not seem to be accidental, as they tend to align with the surface traces of the faults in the region. The combined effects of the carbonates, structures, karstification, and the tectonic history of the area have formed these. It can be said that the cumulative effects of precipitation, charging, and major streams have shaped the water tables we see today. Since these variations are still only on the surface, there is reason to believe that intervention beyond the plain's boundaries is also a factor. However, another factor is crystallized karstic limestones found in the basement. Therefore, water from the limited water horizon offers better potential (Dolines and several obruks Figs. 4.6, 4.7, 4.8, 4.9 and 4.10).

4.4.2 Morphometry of Obruks

Apart from Kızılca, Bellikuyu, and Tahtalı sinkholes, which were developed within the Mesozoic limestones, almost all sinkholes were developed within the Neogene Lake limestones.

These sinkholes are dry. Figures 5.3 and 5.4 shows the illumination image of the Konya Basin sinkhole plateau's digital elevation model. Neogene limestones are generally horizontal and sometimes intercalated with clay and marl.

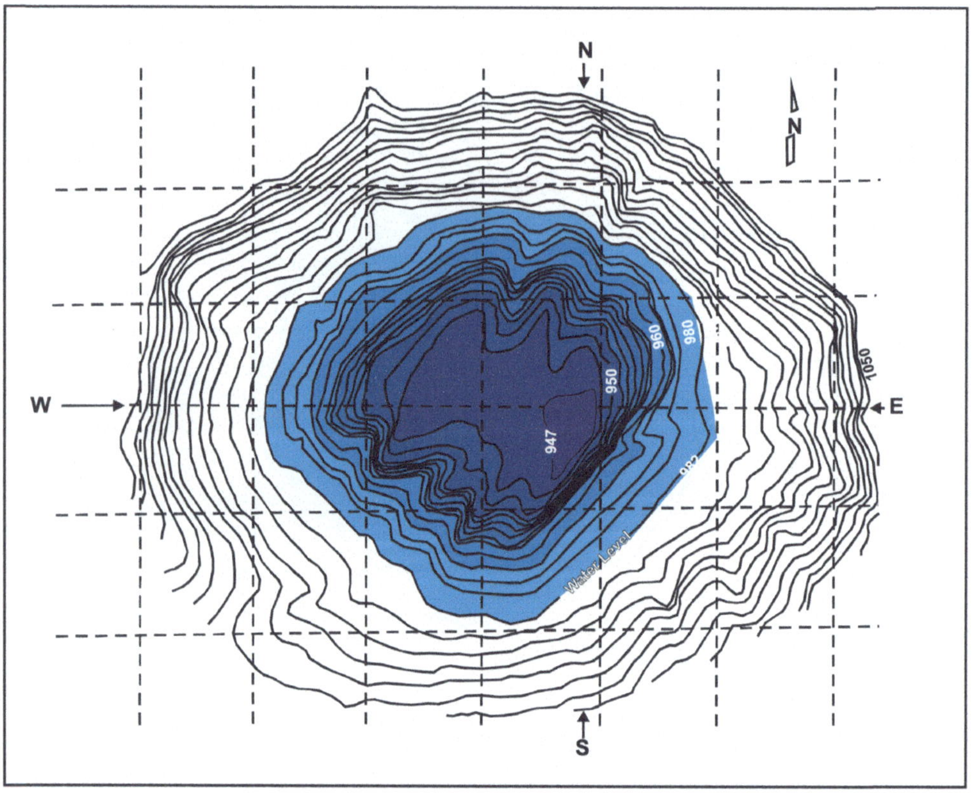

Fig. 4.4 Bathymetric map of the Kizören Obruk (after Çörekçioğlu 1994)

Fig. 4.5 Hydrologic studies at Kizören Obruk area

Fig. 4.6 Dolines in the Konya-Karapinar plain

The obruk development is concentrated in the Obruk Plateau, which is located between Kızören and Karapınar and covers an area of approximately 120 km.

However, some important sinkholes from this area, such as Apasaraycık Obruk and Timraş, developed near the Apa Dam in Neogene conglomerates. Obruk near Çumra developed in connection with the Mesozoic limestones cropping out near the obruk.

4.4.3 Recen Obruks

Akviran Obruk was formed by a collapse in 1977. It is 60 m deep with water at the bottom. It has a conical shape and is only 15 m in diameter at the top and 30–35 m at the bottom.

The wall is made of silt, clay, and marl. Nebili Obruk was developed in 10 m diameter silts and marls in 1983. This obruk expands toward the bottom again. It is 70 m deep and hits the water table.

However, the parts formed due to the collapse and the ensuing landslide cover the water table. Sekizli Obruk area includes more than one obruk that has been recently developed. One of the obruks was created in 1983, starting with a 10–15 cm hole.

It collapsed a few months later due to heavy rains. The lithology in these sinkholes consists of silt, clay, and march. It has a diameter of 5 m and a depth of 8.5 m. Another obruk in Sekizli site occurred in May 1995. It is a huge sinkhole with a diameter of about 45 m and a depth of about 60 m. Lithology is again clay, silt, and marl. Collapsing still continues in this area.

Fig. 4.7 Dolines in the Konya-Karapinar plain

4.4.4 Old Obruks

The most typical and beautiful examples of old obruks are Meyil, Kızören, Çıralı, Tirnraş, Karain, Dikmen, Yunus, Hamam, Kızıl, Half, Cup, Potu, Kangallı, Yılanlı, Zincanca, and Kayalıbaşı sinkholes. The first three obruks reach to the water tables. Bathymetric maps of the Meyil and Kızören sinkholes are given in Figs. 5.3 and 5.4, respectively (after Çörekçioğlu 1994).

These sinkholes are arranged in the NW–SE direction. Their diameters vary between 8 and 700 m. The deepest one has a depth of 80 m. Slope, Dikmen, Karain, Kızıl, Kangallı, and Çıralı sinkholes were developed in the limestone.

Çifteler and White sinkholes were developed in marly, silty, and clayey deposits. The water height at Meyil Obruk has decreased from 34 to 30 m in the last 5 years. Long-term water level and rainfall records at the Meyil Obruk revealed that the water level was greatly affected by precipitation.

The Kızören Obruk is one of the most striking structures found in the Konya Closed Basin.

Kızören Obruk, a collapsible doline, has an approximately elliptical shape with a long axis of 180 m and a short axis of 150 m. The maximum depth of the sinkhole is 145 m from the water surface. Although the water level in the sinkhole fluctuates in winter and summer due to groundwater reloading, the change in water level generally does not exceed 1–2 m.

Apasaraycık Obruk is located between Apasaraycık village and Apa Dam. The sinkhole formed in the valley where the Çarşamba stream flows is elliptical with a long axis of about 132 m and a short axis of 88 m. The depth of the water in the sinkhole reaches 52 m at the deepest point.

The lithological unit in which the sinkhole is formed is a basal conglomerate. The conglomerates in this region are observed in layers ranging from 1 to 25 m in thickness, some of which can reach 50 m in thickness. The main constituents of these densely karstified conglomerates are limestone pebbles interconnected with carbonate cement. The Apasaraycık Sinkhole has the same water level as the dam itself, as it is connected to the Çarşamba stream by a gallery excavated before the dam's construction.

Fig. 4.8 Dolines in the Konya-Karapinar plain

Fig. 4.9 Kizören Obruk at the Karapinar plain

Fig. 4.10 Timraş Obruk in the Karapinar plain

Fig. 4.11 Hydrologic studies at the Çirali Obruk

Fig. 4.12 Hydrologic studies at Kizören Obruk

Fig. 4.13 Dr. Günay in working area

Fig. 4.14 Geological engineer Ilhami Çörekçioğlu. We respectfully remember the value that we have been researching together for many years in the Konya-Karapinar at the "Obruk Plateau"

Due to the temperature measurements made to determine the groundwater movement in the sinkhole, it has been determined that the flow direction is from W to E. This situation shows that the groundwater movement in the sinkhole is compatible with the regional groundwater movement. Since the water quality in the sinkhole is suitable for drinking and domestic use, drinking water is pumped from the sinkhole to the village. The underground water potential of the region is relatively high. Most of the 310×10 m annual groundwater potential of the Konya Plain is provided by this region. Although the annual average precipitation is low (335 mm), the groundwater of the region is substantially recharged by groundwater flow from the Taurus Mountain toward the Salt Lake area (Tuz Gölü).

According to Günay (1995), the paleokarst is responsible for the development of this interesting feature. The Lower and Middle Miocene represents a paleokarst horizon across the country.

The paleokarstic structure and CO_2 development from the old Hasandağı volcano are also effective in the formation of the sinkhole. Carbonic acid, which is formed due to CO_2 in groundwater, dissolves the limestones and creates a porous and hollow structure. Excessive groundwater drawn from wells in the region also affects the formation of sinkholes (obruks).

Karstic features of this period were reintroduced into the hydrological cycle in the plural periods of the Quaternary and became active again. However, the water table of the Neogene aquifer does not match that of the Mesozoic aquifer. Karstification in the Neogene lacustrine limestone was also controlled by changing the water levels of the pluvial lakes in the Konya Basin. However, Günay (1995) attributes an important role to tectonics in karstification. The initiation and widening of the waterway from the surface are due to large faults reaching the limestones underlying the Mesozoic age. The arrangement of the potholes along certain lines shows this effect (Figs. 4.11, 4.12, 4.13 and 4.14).

4.5 Conclusions

Konya-Karapınar "Obruk Plain"'s various hydrogeologic resources have been carried out for many years, and as a result of these, many (approximately 100) drilling wells for irrigation have been drilled. When these works failed after a while, it was decided to transfer irrigation water from other basins to the.

References

Canik B., & Çörekçioğlu İ. (1985). The formation of sinkholes (obruk) between Karapınar and Kızören Konya. In: G. Günay and A. I. Johnson (Eds.), Proceedings of Int. Symp. of Karst Water Res., 193–210 p., IAHS PUBL No:161, Antalya, Turkey.

Çörekçioğlu, İ. (1994). Karapınar kuzeyi obruklar sahası karst hidrojeolojisi incelemesi, DSİ Bölge Md. Konya (in Turkish).

Erol, O. (1985). The relationship between the phases of the development of the Konya-Karapınar obruks and the Pleistocene Tuzgölü and Konya pluvial lakes. In: G. Günay and A. I. Johnson (Eds.), Proceedings of Int. Symp. of Karst Water Res., 207–106 p., IAHS PUBL No:161, Antalya, Turkey.

Eroskay, S. O. (1976). The factors influencing the Konya obruks and their groundwater potentials evaluation, İÜFF Mecmuası seri B., 41, 1–4, 5–14, İstanbul.

Ford., D., & Williams, P. (1989). Karst geomorphology and hydrology, Unwin Hyman, London.

Güldalı, N., & Şaroğlu, F. (1983). Konya yöresi obrukları, TJK Yeryuvarı ve İnsan, cilt 7, sayı 1, Ankara (in Turkish).

Günay, G. (1977). Konya-Sarıcalar dolayının jeolojisi ve yeraltısuyu olanaklarının izotop yöntemlerinden de yararlanılarak incelenmesi, Doktora tezi, İ.U.F.F. Tatbiki Jeoloji Kürsüsü, İstanbul (in Turkish).

Günay, G. (1980). Manavgat havzası ve yakın dolayının karst hidrojeolojisi incelemesi, Doçentlik tezi, Hacettepe Üniversitesi Fen Bilimleri Enstitüsü, Beytepe, Ankara, (in Turkish).

Günay, G. (1995). Guide book, international symposium and field seminar on karst waters and environmental impacts, 10–20 Sept. 1995, Beldibi, Antalya, Turkey.

Günay, G., Çörekçioğlu İ., Eroskay, S. O, & Övül, G. (2010). Konya - Karapınar obruks (sinkholes) of Turkey. In: B. Andreo, F. Carrasco, J. J. Durán and_ J. W. LaMoreaux (Eds.), Advances in Research in Karst Media, 367–372 pp.

Ketin, İ. (1977). Main orogenic events and paleogeographic evolution of Turkey, MTA Dergisi, 88, 1–10 pp., Ankara, Turkey.

Özgül, N. (1984). Stratigraphy and tectonic evolution of the Central Taurus. In: O. Tekeli, M. C. Göncüoğlu, (Eds.), International Symposium on The Geology of the Taurus Belt, 77–90 p., Ankara.

Şengör, A. M. C. (1980). Principles of neo-tectonics of Turkey. TJK publication, 40 p., Ankara.

Origin and Catchment Area of the Köprüçay Karst Springs

5

Mustafa Degirmenci and Gültekin Günay

Abstract

The Olukköprü springs are a significant karst groundwater discharge point in the Western Taurids. Due to this relationship, the Olukköprü springs are of great importance for the Beşkonak Dam project. The Olukköprü springs discharge from Miocene conglomerates at the northern edge of the Beşkonak reservoir and have an average dry-period discharge of 30 m^3/s, whereas, in wet periods, the average discharge can be up to three times higher. Based on measurements carried out at the Beşkonak flow gauging station 4 km downstream from the springs, the long-term average flow of the Köprüçay River is estimated to be 86.4 m^3/s. According to the basin-wide hydrological budget calculations based on the Beşkonak flow gauging station's flow data. An overall analysis of the available data has shown that some 10 m^3/s of this groundwater recharge comes from the Büyük Çandır Basin; the rest might come from northeast of the Köprüçay Basin.

5.1 Introduction

Köprüçay basin is a neighboring basin to the Manavgat basin from the east. The Dumanlı spring is the most important source of this adjacent basin (Günay and Yayan, 1979; Karanjac and Günay, 1980). The Olukköprü karst springs in the Köprüçay Basin are one of Turkey's most important karst discharges and are located within the Köprüçay Canyon National Park, 40 km west of Antalya. These sources are the main water source of the Beşkonak dam planned to be built on the Köprüçay River in the 1960s (Eroskay, 1968; EİEİ, 1973).

It is known that a significant part of the recharge to the springs takes place from adjacent basins. Therefore, in Fig. 5.1, special attention is given to the resources in relation to the water resources projects in the neighboring basins and the Beşkonak Dam project. In this study, the recharge and discharge measurements of these resources were determined and their relations with neighboring basins were examined.

Hydrogeological Processes in Karts Terranes (Proceedings of the Antalya Symposium and Field Seminar, October 1990). IAHS Publ. no. 207, 1993.

M. Degirmenci
Cumhuriyet University, Sivas, Turkey

G. Günay (✉)
Hacettepe University, Yaşamkent Mah. Ataşehir Sitesi 7A-5.Blok No:12, Çankaya, Ankara, Turkey
e-mail: ggunay@ttmail.com

5.2 Geological Units and Hydrogeological Functions

The stratigraphical units in the region can be divided into allochthonous and autochthonous units in terms of their lithological and structural features. Paleozoic, Mesozoic, and Cenozoic rocks prevail in the region.

5.2.1 Autochthonous Units

5.2.1.1 Paleozoic

The Paleozoic rocks consist predominantly of schists and outcrop at the north of the basin (Fig. 5.2). These units have an approximate thickness of 2000 m and serve as the impermeable base in the region.

5.2.1.2 Mesozoic

Triassic shales, known as Kasımlar and Kırkkavak formations (Tr), have a thickness of over 2,000 m and are considered to be the most important impermeable units in terms of their hydrogeological functions. The hydrogeological

G. Günay et al., *Caves and Karst of Turkey - Volume 2*, Cave and Karst Systems of the World,
https://doi.org/10.1007/978-3-030-95361-4_5

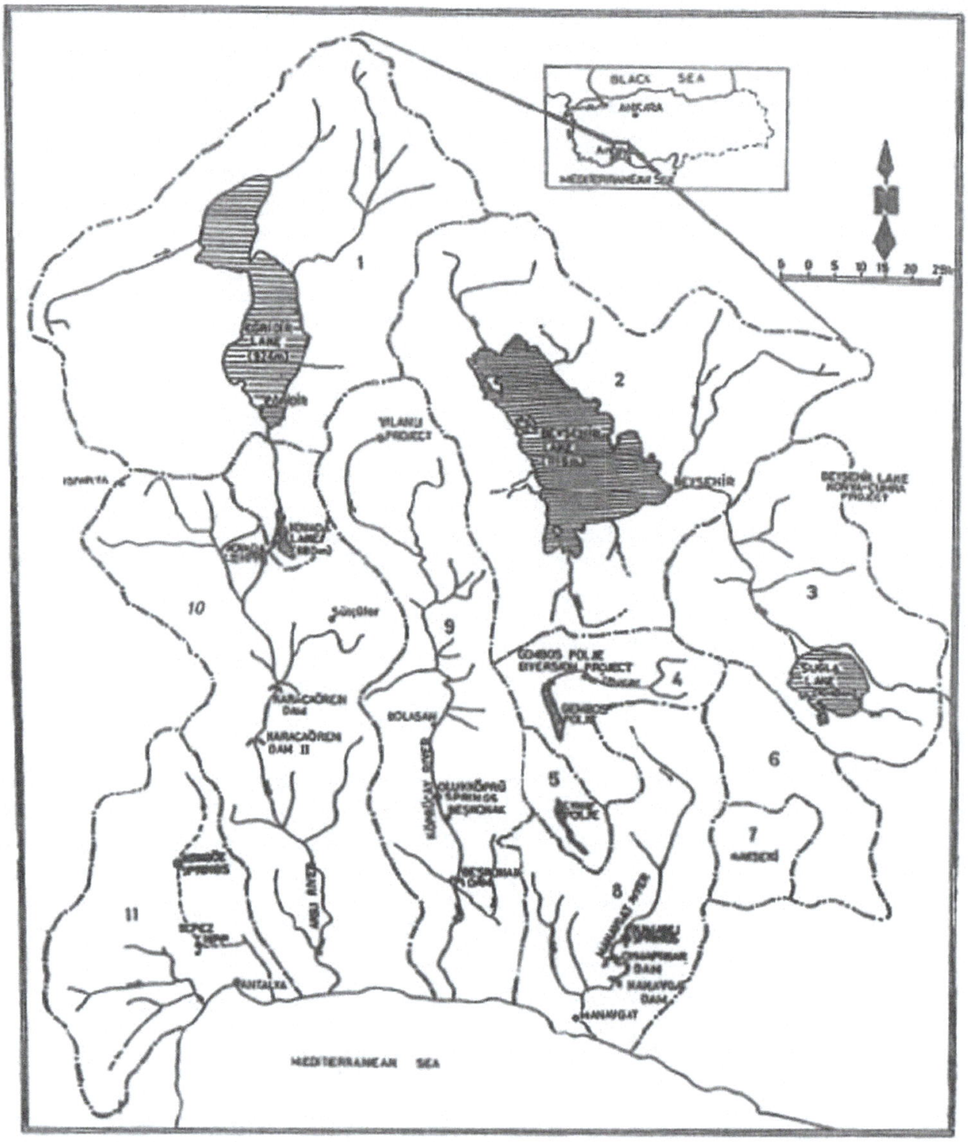

Fig. 5.1 Location map of the water resources projects in adjacent basin. 1—Alluvium; 2—travertines; 3—sandstone-claystone; 4—conglomerates; 5—flysch (sandstone, shale, limestone); 6—limestone; 7—shale, sandstone, conglomerate; 8—limestone, dolomite; 9—schist, quartzite; 10—impermeable beds; 11—limestone; 12—limestone; 13—impermeable beds; 14—main surface water divide; 15—sub-basin surface water divide; 16—underground water divide; 17—probable direction of groundwater flow; 18—groundwater flow direction defined by dye tests; 19—spring; 20—group of springs; 21—river; 22—discharge measurement (A: annual discharge (m^3/s); B: catchment area (km^2)); 23—sinkhole; 24—poljes; 25—cave; 26—hydroelectric power station; 27—dam site; 28—lithological boundary; 29—threat fault; 30—reverse fault; 31—dip and strike of the bed; 32—fault; 33—rose-diagram of joint systems; 34—sampling station; 35—residential area

connections between the Köprüçay basin and the basins to the north and south are closely related to the extent and structure of these units, which constitute natural barriers to karstic groundwater movement; They form a separation between the limestone aquifer of the Anamas Mountains in the north and the permeable units in the south. The discharge from the Anamas Mountains limestone aquifer to the Köprüçay basin occurs through numerous karstic springs such as Başpınar and Karapınar springs, which are located at the contact between the limestone and these impermeable units. Impermeable units also crop out between Kepez and Dumanlı Mountains in the east of Kırkkavak fault, preventing groundwater movement between polje systems and Köprüçay basin.

In the study area, the Jurassic-Cretaceous units are represented by common limestone deposits (J-K) known as the widespread succession. This limestone succession, over 1,000 m thick, is heavily karstified and reveals striking examples of many karstic features such as karens, dolins, uvalas, sinkholes, poljes, caves and estavellers.

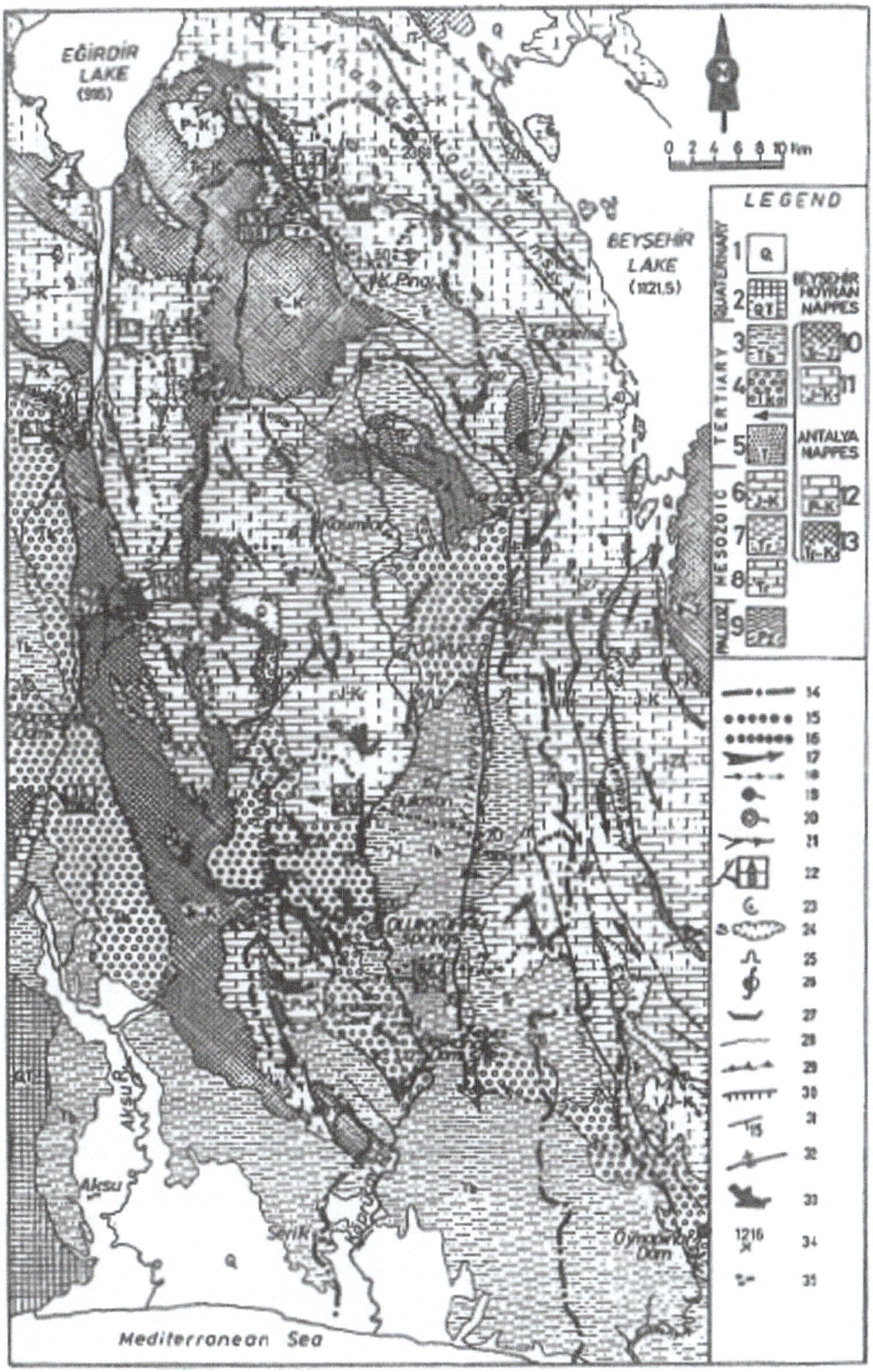

Fig. 5.2 Hydrogeological map of the Köprüçay area

Members of Antalya nappes are Tertiary (T), Lutetian flysch units (Ketin) on the Jurassic-Cretaceous limestones in the Köprüçay and Aksu basins and between these limestones and the Beyşehir-Hoyran-Hadim nappes in the east (Ketin 1977). The impermeable flysch is of particular hydrogeological importance as it acts as an impermeable barrier between the permeable nappe units and the underlying limestones.

Fig. 5.3 Hydrogeologic map of the Olukköprü area

5.2.1.3 Cenozoic

Miocene sediments with a thickness of more than 1000 m are located on the old rock and nappe units in the region. They cover a large area of the Köprüçay basin and comprise gradually alternating conglomerates (Tk, Köprüçay conglomerates) with coarse components and alternating shale-sandstone sequences (Tb, Beşkonak formation). Köprüçay conglomerates are cementitious, cracked, and jointed. There are many karst features in the basin, such as Kuruköprü cave (Değirmenci, 1993).

5.2.1.4 Geological Units and Hydrological Functions

The geological units in the region can be divided into allochthonous and autochthonous units in terms of lithological and structural features. Paleozoic, Mesozoic, and Cenozoic rocks prevail in the region.

5.2.2 Allochthonous Units

Antalya nappe units occur between the Köprüçay Basin and the Aksu Basin to the west of the region. Antalya nappes cover large areas of the Aksu and Köprüçay basins, from Eğirdir Lake in the north to the Serik-Aspendos areas in the south. Permeable and impermeable nappe units are of great importance from the hydrogeological point of view (Fig. 5.2).

5.3 Bulasan and Beşkonak Flow Records

Springs' discharge rates of 30 m^3/s in the year 1964 were observed at the Beşkonak flow gauging station. This flow belongs to the minimum contribution of the Olukköprü, Böğrümköprü, Alabalık pool, and Oğlanuçtuğu springs. The average flow rates for this springs are: Olukköprü springs 31 m^3/s, Böğrümköprü springs 2.5 m^3/s, and Alabalık and Oğlanuçtuğu springs 6 m^3/s. Based on the same flow gauging records, it can be concluded that in dry periods, about 415 of the total flow (48 m^3/s) recorded at the Beşkonak flow gauging station is supplied by the Olukköprü and other springs, whereas 1/5 of the flow comes from the catchment area to the north of the Bulasan spring discharge stations. Based on the budget calculations, about 10 m^3/s of groundwater comes from the Büyük Çandır sub-basin of the Aksu Basin and the remaining 10 m^3/s groundwater inflow to the Köprüçay Basin (Değirmenci, 1989) (Fig. 5.1), (Hydrogeology Figs. 5.2 and 5.3).

5.4 Conclusion

In this study, an attempt was made to determine the hydrogeological connections between the Köprüçay Basin and adjacent basins—thus, the regional karst groundwater flow pattern. For this purpose, a multidisciplinary study was conducted (Değirmenci and Günay 1990, 1993).

References

Değirmenci, M. (1989). Köprüçay ve dolayının (Antalya) karst hidrojeoloji incelemesi, Doktora tezi, Hacettepe Üniversitesi Fen Bilimleri Enstitüsü, Beytepe, Ankara, 398 sf. (in Turkish).

Değirmenci, M. (1993). Karstification at Beşkonak dam site and reservoir area, southern Turkey, Environmental Geology, 22, 111–120 pp.

Değirmenci, M., & Günay, G. (1990). Analysis of hydrologic relations between Eğirdir Beyşehir-Suğla Lakes systems and adjacent basins by means of remote sensing techniques; (South Turkey). Environmental Geology and Water Sciences, USA., vol. 19, No:1, pp. 41–45.

Değirmenci, M., & Günay, G. (1993). Origin and catchment area of Olukköprü karst springs. In: G. Günay, A. I. Johnson and W. Back (Eds.), Hydrogeological processes in karst terrains, Proceedings of the Antalya Symposium and Field Seminar, IAHS Publication No: 207, 97–106.

EİEİ. (1973). Köprüçay-Beşkonak bent yeri ve enjeksiyon perde güzergahları mühendislik jeolojisi incelemesi (Engineering geological investigation of Köprüçay-Beşkonak dam site and injection curtain section) EIE Publ. No. 73-17, Ankara. 86 pp.

Eroskay, S. O., 1968, Köprüçay-Beşkonak rezervuan jeolojik incelemesi (Geological investigation of Köprüçay-Beşkonak reservoir), EIE Report No. II-06-5. Ankara. 80 pp.

Günay, G. & Yayan, T. (1979). Antalya - Kırkgöz kaynakları hidrojeoloji incelemesi. 1.Ulusal Hidrojeoloji Semineri, DSİ Oymapınar Barajı, Antalya. DSİ-UNDP Projesi TUR / 77/ 015 Project Studies. DSİ Groundwater Dept. Yücetepe, Ankara (in Turkish).

Karanjac, J., & Günay, G. (1980). Dumanlı spring Turkey - the largest karstic spring in the world, Journal of Hydrology, 45, p. 219–231.

Ketin, İ. (1977). Main Orogenic Events and Paleogeographic Evolution of Turkey, MTA Dergisi, 88, 1–10 pp., Ankara, Turkey.

6 Tectonic Influences on Groundwater Flow Systems in Karst of the Southwest Taurus Mountains, Turkey

William Back and Gültekin Günay

Abstract

The Taurus Mountains were formed by folding and overthrust faulting during the Alpine orogeny. The epeirogenic coastal movements continued from the late Tertiary into the Quaternary period. The Quaternary tectonism is characterized by normal faults and, in this western part, by block faults. With uplift, karstification increased and the surface drainage of many lakes and rivers was transformed into subsurface drainage. The climate changed drastically several times during the Ice Age. These climatic changes were also controlled on the extensive dissolution of limestone and precipitation of the travertine terraces of Antalya that has a thickness of about 300 m and extends over an area of approximately 630 km^2.

6.1 Introduction

The Karst Commission first met in Turkey in July 1985, during the International Symposium, "Karst Water Resources," which convened in Ankara for the formal presentation of papers and followed with a field trip from Ankara through Konya and Antalya to Izmir. The Proceedings volume was published by the International Association of Hydrological Sciences (Günay and Johnson 1985) and a guidebook was prepared for the field trip. The latest meeting of the Karst Commission was held in connection with the "International Symposium and Field Seminar on Hydrogeologic Processes in Karst Terrains" in Antalya, October 7–17, 1990. Also, at that time, the first organizational meeting of the International Geological Correlation Program (IGCP) Project 299 was held, led by Yuan Daoxian of the Karst Research Institute in Guilin, China. This chapter is expected largely from the guidebooks prepared for those Symposia and from the official report of the IGCP Project 299 (Yuan and Back 1991).

6.2 Tectogenetic Control of Karstification

Karstic limestones in Turkey cover approximately one-third of the country, and they are divided into four main karst regions (Eroskay and Günay, 1980). Among these karst regions, the most important and interesting one is the Taurus Karst region. Karst structures start from the Aegean Sea in the west and are central toward Turkey's eastern border. In general, tectonically, lithologically, and climatologically, they affect groundwater movement. The Taurus Mountains have been significantly influenced by the Alpine Orogeny with folding and thrust faulting (Ketin, 1977). This orogenic system also ultimately significantly affected karst (Fig. 6.1).

The Aegean islands were interconnected, as well as Turkey and Greece at various times. Since the Late Tertiary, Western Turkey has experienced a downward movement largely controlled by faults. Turkey's highly crenelated and fragmented Aegean coastline is due to this tectonic congestion. The epirogenic coastal movements continued from the Late Tertiary into the Quaternary period. Quaternary tectonics is characterized partly by normal faults and partly by block faults.. Many of the coastal bays are in structural grans that extend inland as major valleys. At approximately the middle of the Early Pleistocene epoch, structural activity changed the river systems. In the Tertiary, the river drainage from the highlands was directed toward the north and the south. The Early Pleistocene faulting diverted these rivers

Gültekin Günay: International Contributions to Hydrogeology, Vol. 13 (1992) © Verlag Heinz Heise, Hannover, FRG.

W. Back
U.S. Geological Survey, Water Resources Division 431 National Center, Reston, Virginia, 22092, USA

G. Günay (✉)
Hacettepe University, Yaşamkent Mah. Ataşehir Sitesi 7A-5
Blok No: 12, Çankaya, Ankara, Turkey
e-mail: ggunay@ttmail.com; ; gultekin.gunay6@gmail.com

G. Günay et al., *Caves and Karst of Turkey - Volume 2*, Cave and Karst Systems of the World,
https://doi.org/10.1007/978-3-030-95361-4_6

Fig. 6.1 Many deep canyons such as Dalaman river are formed by dissolution of the limestone along faults and fractures (Photo: Aleksandr Martyanov; https://mapio.net/images-p/16598734.jpg)

into a westerly direction. With uplift, karstification increased and the surface drainage of many lakes and rivers was transformed into subsurface drainage. The carbonate rock units are about 200 km wide along the Taurus Mountains which attain elevations of 2500 m (Günay, 1980). These high mountain ranges, sharp peaks, deep valleys, and narrow gorges cause extremely rugged topography (Fig. 6.1).

The Taurus Mountains in this region contain many very special karst features. At the end of the last Ice Age, approximately 10,000 years ago, the temperature rose rapidly to its present value. As evaporation increased, groundwater levels declined and many springs ceased to flow. Freshwater lakes either turned into saline lakes or dried up entirely. The climate did not change much during the Holocene epoch, with the exception of the period of 5500–2500 BC, when it was somewhat warmer and more humid (Brinkmann 1976). At the beginning of the Holocene, the people of Turkey changed from hunting and gathering society to a planned agricultural society with farming and ranching. Even though the tectonic subsidence is continuing today, new land is being formed by the deposition of sediments in many of the estuaries. The amount of sediments carried by the rivers has increased during the last millennia as a result of deforestation and agricultural practices. For example, the Menderes River (Fig. 6.3) near Izmir has an average delta advance of about 6 m/a (Brinkmann 1976) (Figs. 6.2 and 6.3).

6.3 The Effect of the Quaternary Ice Age Term on Taurus Karst Processes

During the Ice Age, the climate changed drastically several times. Although the Taurus Mountains are without glaciers in the present situation, they were mostly covered by glaciers prior to this period. Throughout Turkey, the snowline during

Fig. 6.2 View of the crenulated shoreline. It is forming many bays where the tectonic structures forming the Taurus mountains plunge into the Mediterranean Sea (Photo: Acar54,https://commons.wikimedia.org/wiki/File:Kelebekler_Vadisi.jpg)

the first global epic was 1000–1200 m lower than now. Thus, the temperature In July would be 6–8 °C lower than today.

The region below the snowline was affected by the glacial climate from the Ice Age, and climatic fluctuations were effective at low elevations. During the glacial epochs, the lakes expanded due to reduced evaporation (Fig. 6.3).

6.4 The Major Controls on the Karst Hydrogeology

Much of the limestone, which ranges in age from Mesozoic to Holocene, has either been overthrust or deposited on formations with extremely low permeability. The combination of this extensive impermeable lower boundary of the regional karst aquifers along with the numerous faults associated with the tectonism is the major control on the karst hydrogeology of this region. The Taurus Mountain region is 4the world's largest karst aquifers, and the largest karst springs. These springs of exceedingly large discharge are formed by the channelized flow along with the fractures by water that cannot penetrate into the deeper formations because of their low permeability; many of the large springs occur at this contact. The original heterogeneity of the aquifer formed by the tectonism has been increased by the dissolution of the limestone. This dissolution results from infiltration of the great amount of rain and snow that occur in the Taurus Mountains, particularly at the higher elevations. If the transmissivity of the carbonate aquifers were more homogeneous, and the aquifers were hydrologically connected to the underlying sediments, storativity of the entire system would greatly increase, and diffuse flow would occur.

This area is characterized by abundant dolines, large poljes, coastal and submarine springs, sea caves, and large travertine terraces. One of the most unusual types of karst is that formed in an extensive conglomerate. The conglomerate is composed of cobbles of limestone with calcareous cement. The mode of transportation and deposition is not well understood. Participants on the field seminars had the opportunity to observe many of these features. Some of the

Fig. 6.3 View of the heavily braded Menderes river. The sediments are derived from erosion caused in part by agricultural activities during the past few centuries. The Menderes river (earlier spelling Maiandros) is the origin of the term "meander".

major conclusions of the participants can be summarized as follows (Yuan and Back 1991) (Fig. 6.4):

1. The formation of karst in southwestern Turkey is strongly influenced by the alternation of Jurassic and Cretaceous limestones with ophiolites and Tertiary flysch in both the horizontal and vertical planes, resulting from nappe structures. Many large springs, some of which are submarine, discharge through structural windows where carbonate rocks are exposed within ophiolite.
2. The general characteristics of this region reflect a combination of both humid and arid environments, a combination that does not occur in tropical climates. This unusual combination in Turkey results from the Mediterranean climate which is characterized by dry summers, wet winters, and a long period of snow cover in the mountains. Therefore, mechanical weathering is as important as chemical corrosion.
3. Because of lower water temperatures, the coastal karst of this area displays effects of biogenic processes less than does that of the Caribbean or of the southeast Asia region.
4. The hydrogeologic conditions of the karst indicate two types of recharge:
   a. Systems that are recharged by poljes on lower plateaus are characterized by high fluctuations of discharge, higher temperature, higher bicarbonate content, and extensive travertine deposition;
   b. Systems recharged mainly from high mountains that are covered by snow for long periods are characterized by more stable regimes of discharge, lower temperatures, lower content of bicarbonate, and little or no travertine deposition. (Fig. 6.5).

6.5 Antalya Travertine Plateau

The Antalya travertine deposits have formed as a result of the precipitation of carbonate minerals due to the outgassing of carbon dioxide from water initially saturated with respect to the carbonate minerals (Herman and Hubbard 1990). The source of calcium and carbonate ions is the limestones of the Taurus Mountains in the

Fig. 6.4 Prof. Dr. Yuan Daoxian, Leader for IGCP project 299

north of the plain. The travertine in this plateau has a thickness of about 300 m and extends over an area of approximately 630 km^2. These deposits exist as three separate terraces; one at the altitude of −300 m, another between 50 and 150 m, and the third terrace is below sea level (Fig. 6.5).

6.6 Hydrothermal Pamukkale Travertine

Another major tourist attraction in the area of the field trip of hydrogeologic significance is the spectacular travertine deposit of Pamukkale (Figs. 6.6 and 6.7). Pamukkale means "cotton castle" and is on the ancient site of Hierapolis which was a major city even before the Romans came. The travertine terraces are deposited from the series of geothermal springs that are brought to the surface along faults bordering the graben that forms the valley. The area has some potential for geothermal energy.

The source of the major thermal spring for the travertine terraces and ponds is a cave that had long been known as a road to Hades. The water from the thermal spring is 35 °C, and originally the spring was used for the medicinal qualities of the water. The atmosphere in the cave could not support life, probably due to the high concentration of carbon dioxide gas and consequent lack of oxygen. The reputation of the city was further enhanced by the miracles that the priests would perform by using the cave. It was well known that the cave was dangerous, yet the priests could stay in for long periods, presumably by following the pathways where the upper chambers contained adequate oxygen, then emerge with no adverse reactions. That would then give them the aura of immortality and they could then exert their will on the people. In the second century AD, the Romans diverted the warm mineral water into the buildings for the baths. The Roman bathhouse is now a museum and the thermal springs have been made available to tourists through large swimming pools associated with a motel (Figs. 6.6 and 6.7).

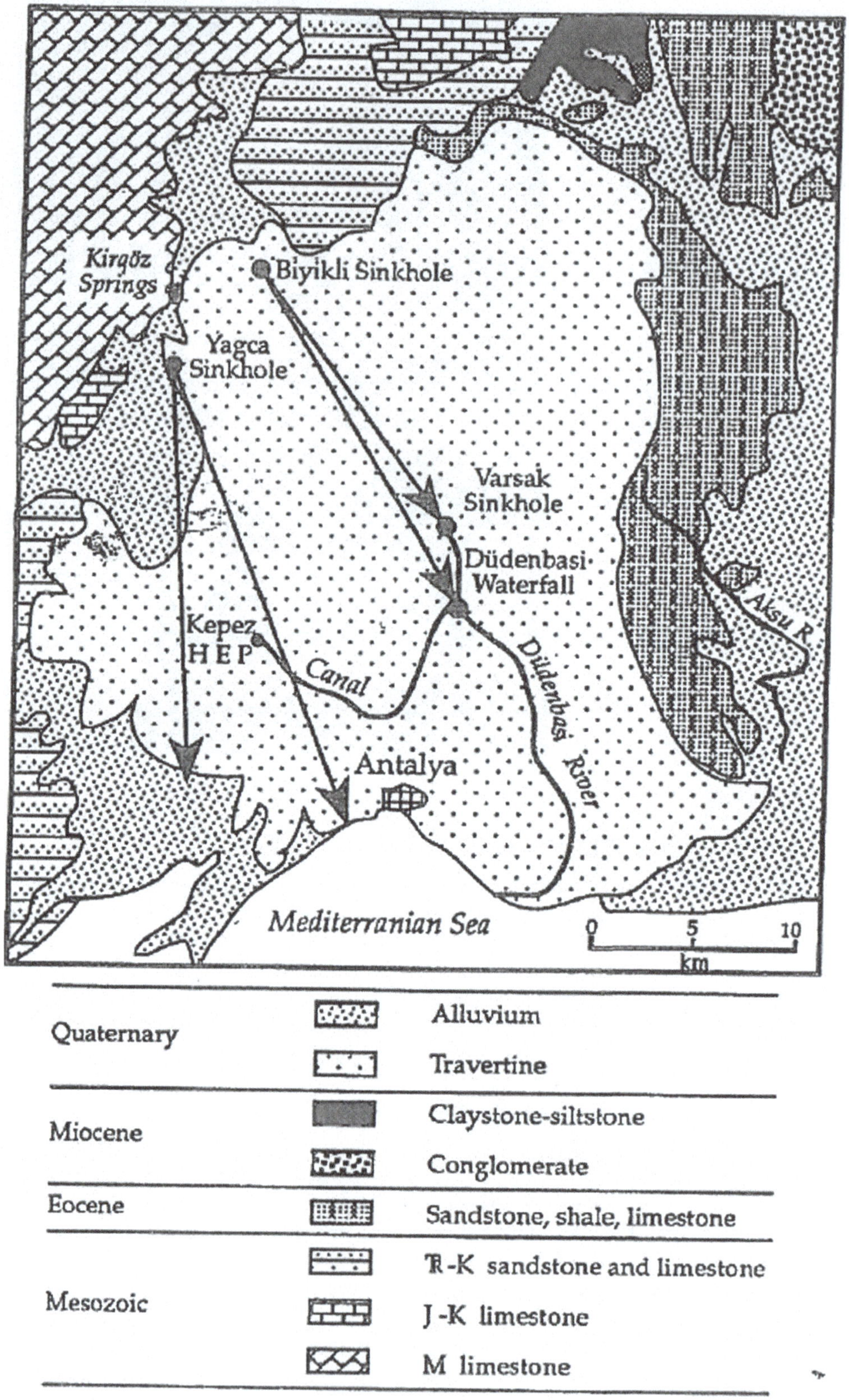

Fig. 6.5 Map showing geologic control on the flow of surface and ground water in Travertine terraces of Antalya (Back and Günay, 1992)

Fig. 6.6 Travertine terraces of Pamukkale, with ruins of the ancient city of Hierapolis and the Roman baths in the backround (Photo: Antoine Taveneaux,https://upload.wikimedia.org/wikipedia/commons/d/de/Pamukkale_30.jpg)

Fig. 6.7 Close-up of the Pamukkale terraces containing rimmed ponds of warm water

References

Back, W., & Günay, G. (1992) Tectonic Influences on groundwater flow systems in karst of the southwest Taurus mountains, Turkey, in W. Back, J. S. Herman, H. Paloc (Eds.), Hydrogeology of Selected Karst Regions, International Contributions to Hydrogeology, Vol. 13, pp. 263–272, Verlag Heinz Heise, Hannover, FRG.

Brinkmann, R. (1976). Geology of Turkey: Ferdinand Enke Verlag, Stuttgart, 158 p.

Daoxian, Y., & Back, W. (1991). IGCP Project 299: Geology, climate, hydrology, and karst formation: Episodes, v. 14, no. 1, p. 80–81.

Eroskay, S. O., & Günay, G. (1980). Tecto-genetic classification and hydrogeological properties of the karst regions in Turkey. in G. Günay (Ed.), Karst Hydrogeology Proceedings: October 1979, Oymapınar-Antalya, Turkey, UNDP Project TUR/77/015, pp. 1–41.

Günay, G. (1980). Manavgat havzası ve yakın dolayının karst hidrojeolojisi incelemesi, Doçentlik tezi, Hacettepe Üniversitesi Fen Bilimleri Enstitüsü, Beytepe, Ankara, (in Turkish).

Günay, G., & Johnson, A. I., eds. (1985). Karst Water Resources: IAHS Publication No. 161, 642 p.

Herman, J. S., & Hubbard, D. A. Jr. (1990). Travertine-marl: Stream deposits in Virginia: Virginia, Virginia Division of Mineral Resources, 184 p.

Ketin, İ. (1977). Main orogenic events and paleogeographic evolution of Turkey, MTA Dergisi, 88, pp. 1–10, Ankara, Turkey.

7 Karst Areas of Turkey

Eric Gilli

Abstract

Turkey offers a great diversity of karst areas that extend from high mountains zones to the Mediterranean and Black Sea shores. They concern limestone and dolomite series but also gypsum deposits. In addition to normal karst features, Turkey contains several forms of global interest: tufa deposits, travertine and surface rimstones, large poljes, and numerous obruks, a kind of collapse sinkholes. There is no important horizontal cave but very deep shafts are encountered in the Taurus zone. The important variability of Turkish karsts is related to its active tectonics and to the sea-level variations in the coastal regions.

7.1 Introduction

The high diversity of Turkey's landscapes results from a complex geologic history during which present landforms have been shaped by intense tectonics related to the collision between the African, Arabian, and Eurasian plates since the mid-Miocene (Fig. 7.1).

To the north and south, the Turkish relief includes mountain ranges that plunge into the surrounding seas (Mediterranean, Aegean, Marmara, Black Sea) leaving only narrow coastal zones. Between both mountains, the vast plateau of Central Anatolia extends. It rises gradually to the east, toward the Anatolian highlands. Several high volcanoes are present inland, among which is the Ararat (5136 m) which is the highest summit of Turkey. The mean altitude of the country is quite high (1100 m).

The climate of the western part is mainly of Mediterranean type, with mild winters and hot and dry summers. The Anatolian Plateau which is surrounded by high mountains offers a more continental climate, dry, cold in winter, and very warm in summer, while in the northern zone, close to the Black Sea, it is more humid.

Lithology also explains the variability of the relief and the diversity of the karst zones. Soluble rocks concern 40% of Turkey's superficy and karsts are present in the whole country, from high mountain zones like Akdağ Mountain (3016 m) to sea shores. The ages of the rocks vary from the Paleozoic to late Cenozoic. In some places, the total thickness of the main carbonate series may overpass 1500 m.

Surface features are well developed in several zones like the Taurus Mountains, the Konya area, or the Black Sea Mountains. Some specific forms like the Antalya tufas, the Pamukkale rimstones, the Taurus poljes, and the Konya obruks are known worldwide. Dumanlı spring and the watershed of Ras-al-Aïn (Syria), that are two of the ten most important karst springs in the world, are located in the south of Turkey which emphasizes the importance of karst aquifers for Turkey's hydrogeology.

Karsts and cave studies started in the 1960s with the work of Temuçin Aygen a geologist from Ankara University. In the 1970s Turkey undertook an ambitious development program for hydroelectricity and irrigation, consisting of several large dams like Keban, Atatürk, and Oymapınar. They encountered difficulties caused by the presence of karst systems, which are also responsible for important leaks. In addition to the problems in land management and water supply in the karst area, they triggered an applied research.

Fundamental karstology studies are recent, and research fields are numerous which makes Turkey a prime location for karst studies. The present chapter gives a short overview of Turkey's karsts. It mainly focuses on the places where caves are present. More information on karst features, and currently explored research fields, is available in an important synthesis that was done by Kuzucuoğlu et al. (2019).

E. Gilli (✉)
Paris 8 University, Saint-Denis, France
e-mail: eric.gilli@univ-paris8.fr

G. Günay et al., *Caves and Karst of Turkey - Volume 2*, Cave and Karst Systems of the World,
https://doi.org/10.1007/978-3-030-95361-4_7

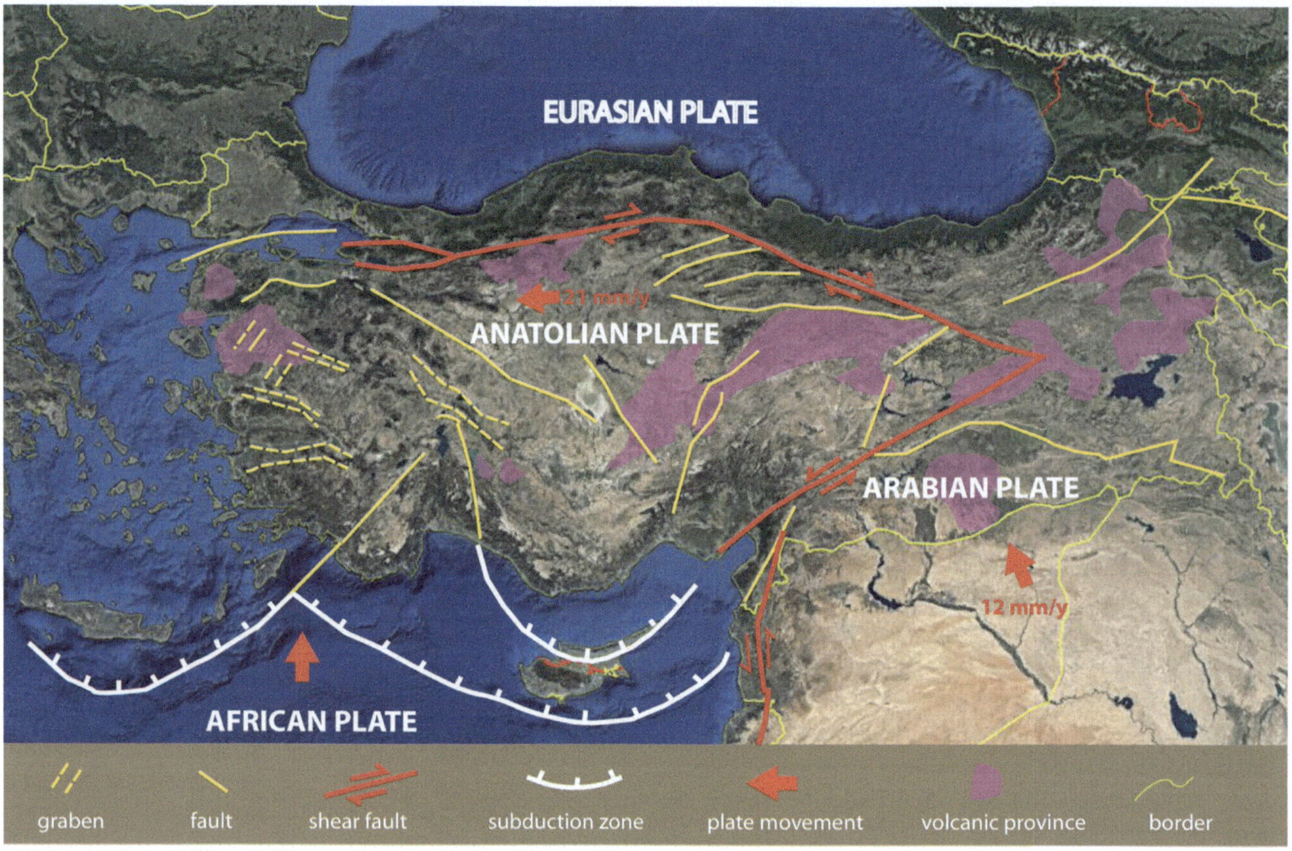

Fig. 7.1 Tectonics and volcanic map of Turkey

7.2 Karst Limestone

Six limestone or dolomite karst regions are classically described that concern whole Turkey (Nazik et al. 2019) (Fig. 7.2):

- The Taurus Mountain;
- The Thrace and Black Sea Mountains;
- The Western Anatolia;
- The Central Anatolia;
- The Eastern Anatolia;
- The South Eastern Anatolia.

The first two are the most important ones. They are located in the northern and southern parts of Turkey. Karst landforms and caves are well developed.

7.2.1 The Taurus Mountains

The Taurus Mountain forms a high ridge on the southern coast of Turkey and extends to the Iranian border. It is affected by important neotectonics movements and the variations of the Mediterranean Sea level for its western part. That zone contains several large poljes, and the longest and deepest caves of Turkey: Pinargozu Cave, (8500 m long) near the Beyşehir Lake and the Peynirlikönü Cave (1492 m deep) in south of Anamur.

In some places lapies are intensely developed like in the Giden Gelmez Mountain whose name means "the place where one goes and never comes back." Indeed, the absence of water and the presence of sharp and steep karrens, that are sometimes tens of meters high, make human travel extremely difficult. It is the karst zone, where most of the cave explorations took place. Several rivers drain the water toward the south and powerful karst springs are present like Dumanlı in the Manavgat River or Kırkgözler in north of Antalya. The latter is responsible for thick deposits of tufa.

7.2.2 Thrace and the Black Sea Mountains

This zone extends in the northern part of Turkey. Kızılelma (6630 m long), which is the fourth longest cave of Turkey is in that area, close to Zonguldak. This ridge is located in the north of the Northern Anatolian Fault, but it is not affected by important neotectonics movements. The development of

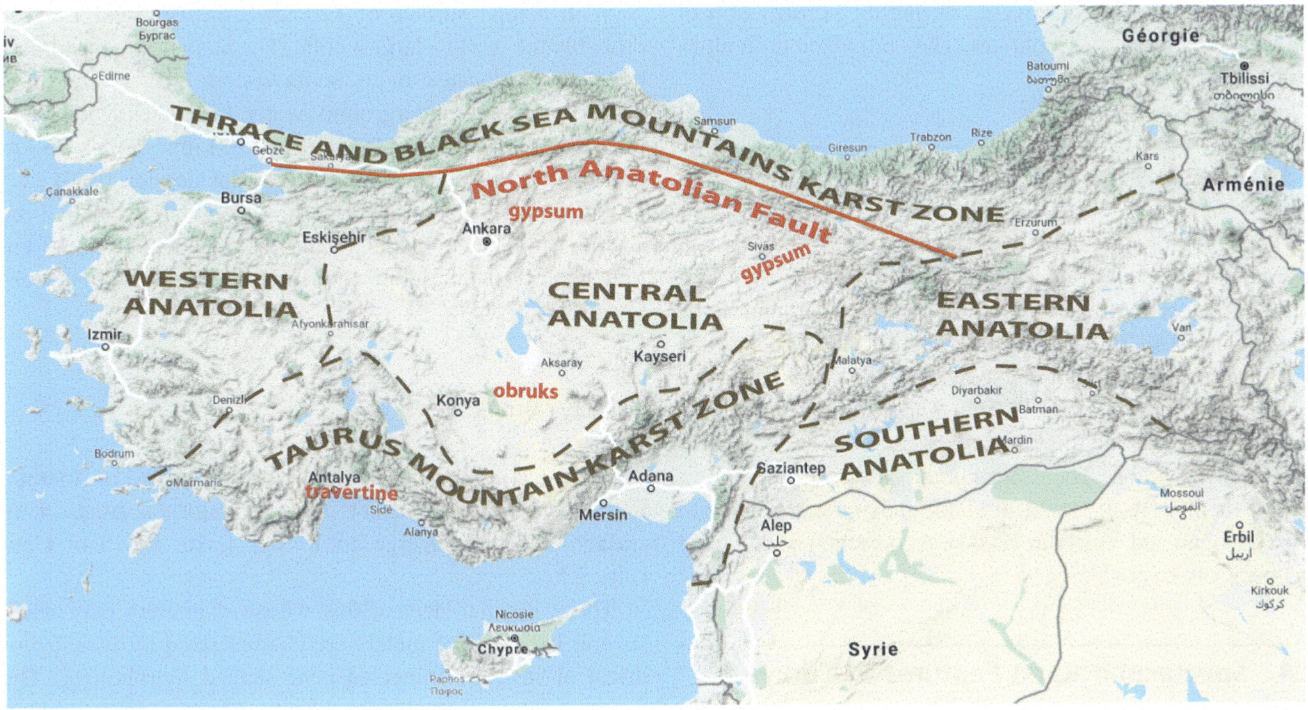

Fig. 7.2 Different karst zones of Turkey

karts systems is mainly directed by the variations of the Black Sea level that cause valleys to deepen during the glacial stages (Nazik and Tuncer 2010). Submarine karst springs are probably present.

7.2.3 The Western Anatolia

This part is bordered by the Aegean and Marmara seas. The rock series that include marble, and the carbonate series from the Permian to Cretaceous are dissected by extensional tectonics, and by a fluvial dissection controlled by the sea-level variations. Although the karst systems are not well developed, two important caves are present: the Ayvaini cave (4866 m long) near Bursa and the Yazören cave near Balikesir.

7.2.4 The Central Anatolia

This zone forms a depression between the Taurus and Black Mountains ranges. Its southern part is an endorheic basin, the Konya plain, where water is drained toward a Salt Lake, the Tuz Gölü. Fossil or current karst features and cave systems are present in several limestone formations, but also in the Oligocene evaporites. The longest cave of that zone is the Kuzgun sinkhole (3187 m long), close to Niğde.

In east of Konya, the Obruk plateau is a Miocene to Pliocene lacustrine limestone unit, where numerous collapse sinkholes, the obruks, are present. They often offer a window on the upper aquifer that extends in these series. The drainage of the karst aquifers of the Central Anatolia is toward the Tuz Gölü or toward valleys that cross the Taurus and Black Sea Mountains (Bayari et al. 2009), but in the Obruk plateau, a vertical drainage is likely, toward a very deep karst system in the Mesozoic series, and then to the Taurus karst springs.

7.2.5 The Eastern Anatolia

This region is a volcanic zone where tectonics has been active since the middle Miocene (Schildgen et al. 2014). Due to the importance of insoluble rocks and the deepening of the valleys that dissect the limestone massifs, karsts systems are not well developed. Cave exploration in that area did not make it possible to discover important caves.

7.2.6 The South Eastern Anatolia

It extends to the north of the Syrian and Iraqi borders. It is part of the Arabian plate. It is a wide Tertiary limestone plateau with classical karst features. Part of the plateau drains to the Ras al-Ain spring in Syria. In Elazig province, important caves were discovered during the construction of the Keban Dam on the Euphrates. They provoked difficulties to fix leaks (Gilli 2015).

7.3 Gypsum Karst

Turkey contains one of the world's largest gypsum karst. It extends in the Çankırı and Sivas provinces in Central Anatolia. The most significant outcrops are located around Sivas (Doğan and Özel 2005). They offer a great variety of surface karst features. An important synthesis work was done by Doğan and Yesilyurt (2019). The evaporites are dated back to the Oligocene. The thickness of gypsum reaches 500 m in the Acıçay Canyon. There are numerous caves, but only a few of them were explored in Hafik and Imranlı (Waltham 2002), Doğan and Yeşilyurt (2004). At present, no long cave has been discovered.

7.4 Spectacular Karst Features of Turkey

Turkey offers specific karst landforms that deserve to be described more precisely.

7.4.1 Large Poljes

In the western part of the Taurus Mountain, several important poljes are present. They are either dry like the Eynif and Kembos plains (Doğan et al, 2017) or filled with water like the Beysehir or Eğirdir lakes (Fig. 7.3).

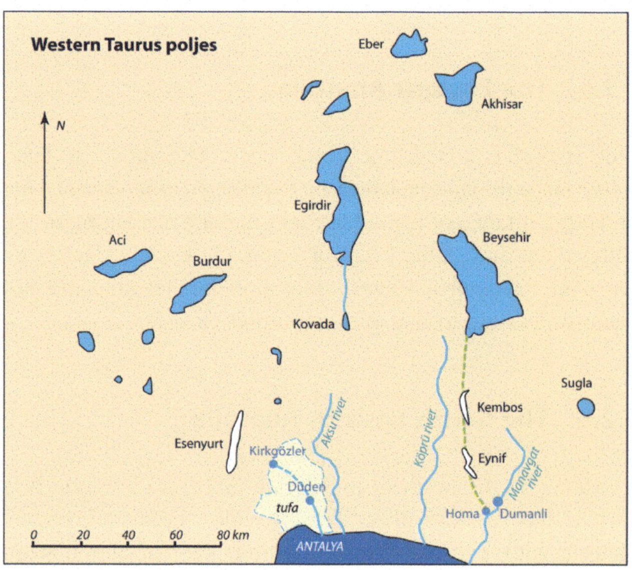

Fig. 7.3 Poljes of Western Taurus

With a total surface of 120 km^2, the Beysehir Lake is probably the world's largest polje. The main ponor that drains the water is located in its southwestern corner. Dye tests proved that the underground water circulates within the limestone of the Taurus Mountains until Homa springs, 75 km to the south, in the Manavgat valley (Bakalowicz 1970).

Eğirdir is 50 km long and is the second largest lake of the Taurus Mountains. To the south, it is connected by a 20 km long valley to the small Kovada Lake, where ponors are located (Fig. 7.4).

7.4.2 Obruks

Obruk is a specific form of sinkhole encountered in Anatolia in the Konya Closed Basin (KCB), an endorheic zone, where surface waters converge to a Salt Lake, the Tuz Gölu (Fig. 7.5).

These circular depressions whose diameters may reach several hundreds of meters are characterized by their cylindrical or inverted truncated cone-shaped morphology. The depth is more than 100 m like in Ciralı (145 m), Meyil (166 m), or Kızoren (225 m) (Fig. 7.6) obruks. They are numerous in the Obruks plateau (Fig. 7.7). Some obruks have a volcanic origin like Meke Maar, but most of them are developed in lacustrine Neogene carbonates. Some of them contain lakes that are windows on the water table. They are used for water supply but overexploitation of the resource causes them to dry up.

During some decades, tens of new sinkholes occurred (Fig. 7.8). Like most of the places, the drop of the groundwater table due to intense pumping for agriculture is the main cause of the recent obruks (Doğan and Yılmaz 2011).

For the older obruks, different genesis processes were assumed:

- explosive volcanic origin (Frech 1916);
- enhanced dissolution caused by high stands of the pluvial lakes during the Pleistocene wet periods (Erol 1986);
- enhanced subsurface dissolution mechanism driven by volcanogenic gas discharges (Canik and Çörekçioğlu 1986);
- hypogenic karstification and mixing corrosion between surface recharge and groundwater (Pekkan 2004);
- dissolution of salt diapirs in a deep thermal aquifer (Arikan 1985);
- upward migration of a deep-seated carbon dioxide flux from an intrusive magmatic body into the Neogene aquifer (Bayarı et al. 2009a).

Another possible explanation that has not been explored yet is the possible dissolution of deep masses of gypsum.

Fig. 7.4 Kovada Polje south of Eğirdir

Indeed, according to Bayarı et al. (2009b), there are two main aquifers within the KCB, which are separated by a Paleocene aquitard. The lower aquifer is in the Paleozoic–Mesozoic series that contain limestone. It could be drained toward the south or the southwest (Fig. 7.5). The upper aquifer is in the Neogene series in which limestone controls the formation, and the areal distribution of sinkholes. It also contains layers or masses of Oligocene gypsum (Çemen et al. 1999). Vertical drainage of the upper aquifer through the aquitard or through a faulted zone could provoke the dissolution of the gypsum, the collapse of deep caves, and the formation of the obruks.

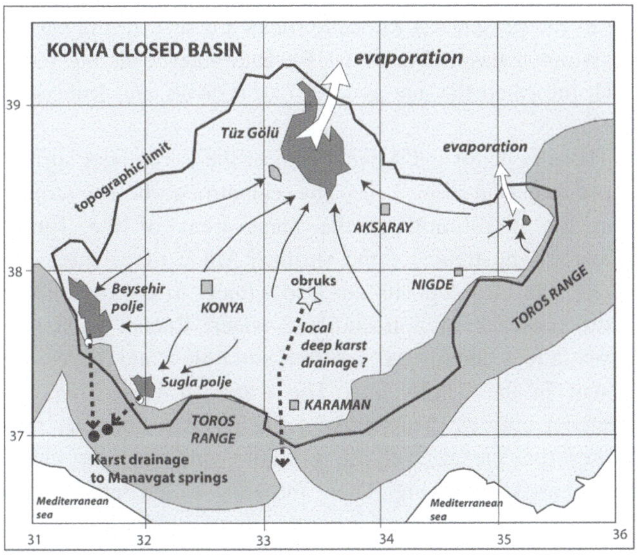

Fig. 7.5 Regional drainage of Konya closed basin

Different observations support that hypothesis:

- a piezometric depression in the water table in the obruks zone (Canik and Çörekçioğlu 1986) (Fig. 7.9) that suggests a vertical drainage of the surface aquifer toward a deeper one;
- close or imbricated obruks that suggest that after the first collapse, the lateral dissolution of the gypsum continues and provokes other collapses (Fig. 7.7);
- apparition of new obruks provoked by pumping for water supply (Fig. 7.8) that shows that instability is close to the surface.

Fig. 7.6 The Kizören Obruk and its pumping plant, 60 km ne of Konya (Photo R. Straub)

7.4.3 Tufa and Travertine Deposits

Important terraces of tufa are present in Antalya and the whole city is built on them. This formation is frequently misnamed as Antalya travertine, but the genesis does not involve the hot water, which makes the term tufa more suitable. These tufa deposits are considered as the largest ones in the Mediterranean countries.

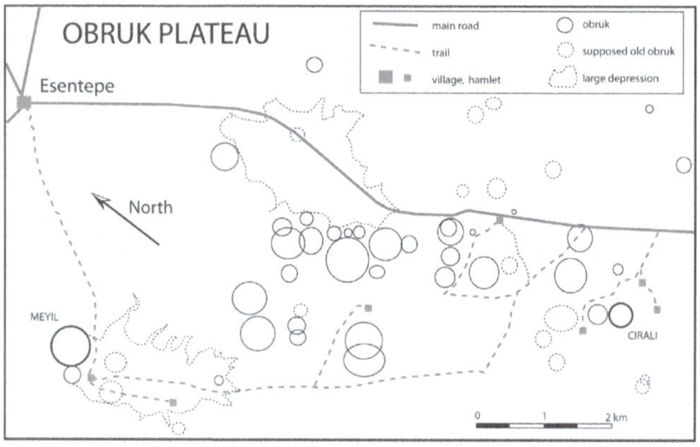

Fig. 7.7 Position of the sinkholes of Obruk plateau

They extend from the Taurus Mountains to the Mediterranean Sea for 630 km^2 between the altitude of 300 m and the depth of 50 m below the sea level. They form several steps in the landscape. The maximum thickness is 275 m (Öziş 1992). The formation probably started in the Pliocene age to the present, as Pliocene rocks are seen in the east of the Antalya basin (Penck 1918). Radiometric dating using U/Th indicated the age as > 600 ka (Glover and Roberston 2003).

The tufa deposited downstream of the Kirkgözler springs is presented in Fig. 7.10. The calcium carbonate comes from the dissolution of the karst areas of the Taurus Mountains upstream of the springs. Made of calcium carbonate, the tufa deposits are also subject to the dissolution in the presence of non-saturated water. Recent karst processes affect them, and several sinkholes and caves are present in the whole area. The water of Kırkgözler and Pınarbaşı springs disappears in the Bıyıklı sinkhole, then crosses the Varsak collapse sinkhole, and emerges in the Düdenbası karst spring. Then, the karst water deposits tufa again as they reach the cliffs that overlook the sea (Fig. 7.11). The water that cascades over the recent tufa deposits forms a very picturesque spectacle.

Fig. 7.8 Recent Obruk in Konya plain. May 17th, 2015 (From Euro News)

The second interesting but smaller place is Pamukkale ("The cotton castle" in Turkish). It is on the list of natural and cultural World Heritage and also includes the antique city of Hierapolis. In Pamukkale, abundant white deposits of travertine form picturesque rimstones (Fig. 7.12). They are formed around geothermal springs related to an active fault (Altunel and Hancock 1993). These deposits are the result of deep dissolution within thick limestone series crossed by thermal water (Altunel and D'Andria 2019). It is more than 400,000 years old (Altunel 1994). The chemical composition of the geothermal shows that it results from the mixture of normal karst water and warm water, upwelling from deep hydrothermal sources (Lu et al. 2016). The rimstones of Pamukkale are the testimony of a hypogene karst, currently in evolution. Similar formations are also visible in Akkayalar (Bolu in north-eastern Turkey).

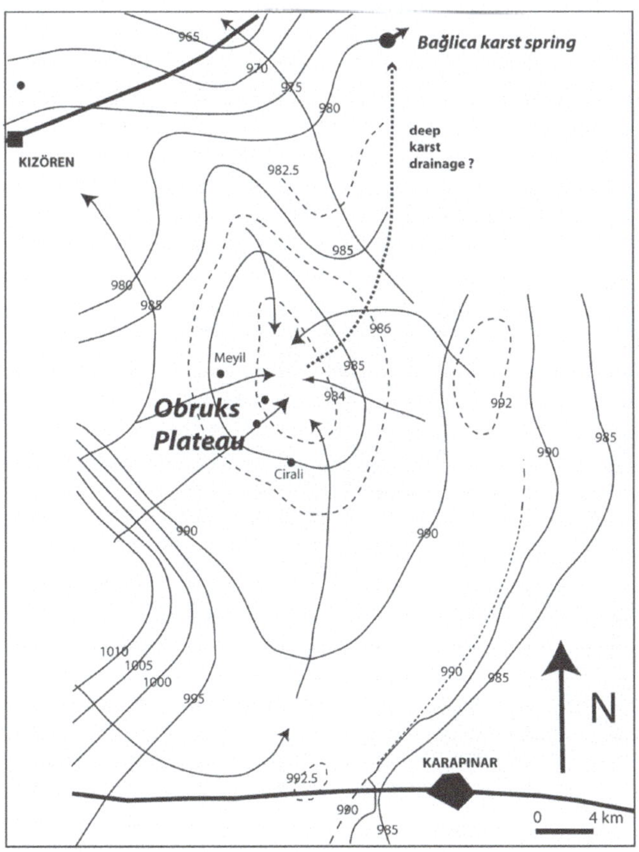

Fig. 7.9 Piezometric depression in the Obruk plateau after Canik and Çörekçioğlu (1985)

7.5 Karst Hydrogeology

Karst covers 40% of Turkey, thus karst hydrogeology is an important domain. Even if we do not encounter new discoveries of long cave systems in Turkey, the importance of karst aquifers shows that some karst systems are huge. Close to Manavgat in the southern part of the Western Taurus, Dumanlı (Manavgat) is one of the ten world's largest karst springs (Fig. 7.13). The average discharge is 50 m^3/s, and it overpasses 1000 m^3/s during flood periods. It is now below the surface of the reservoir lake of the Oymapınar dam on the Manavgat River. The karst aquifer and its watershed remain unknown.

Ras al-Ain (35 m^3/s) is also one of the most important karst springs in the world. The outlets are located in Syria but half of the watershed extends in Turkey (Nicod, 2000) between the cities of Mardin and Urfa (Burdon and Safadi 1963; Cater et al. 1994). There is no information of any cave system related to that aquifer. In recent years, the withdrawal of groundwater for irrigation led to a dramatic decrease in spring discharge, and Ras al-Ain has practically dried up (UNESCWA and BGR 2013).

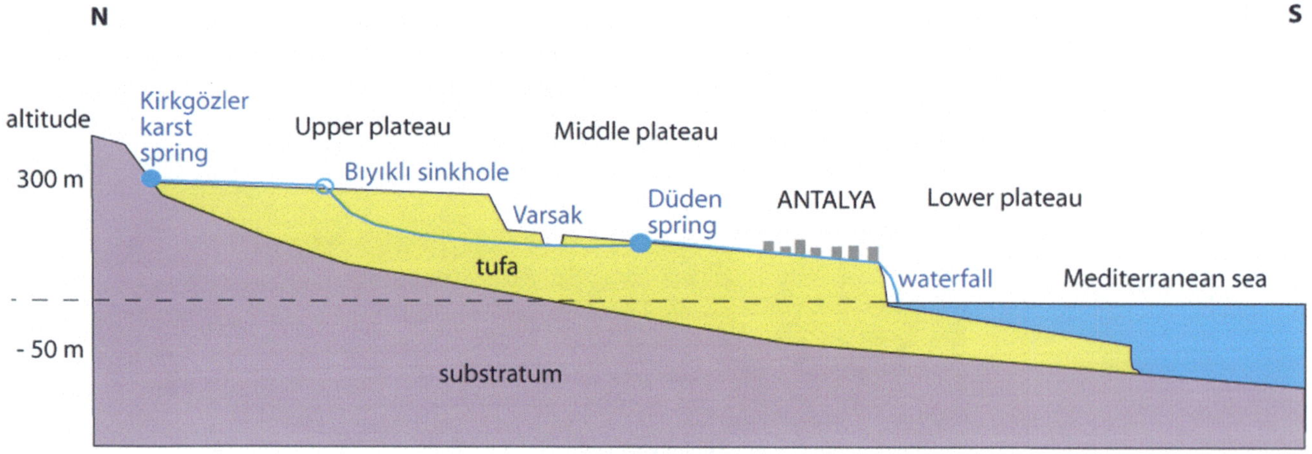

Fig. 7.10 Tufa deposits in Antalya region

Fig. 7.11 Waterfall and tufa deposit in Antalya

Fig. 7.12 The surface rimstones of Pamukkale (Photo ISMET11 - Wikimedia Creative Commons)

Since the 1970s, important hydraulic work was done and a great number of dams were built. They sometimes encountered difficulties in karst areas. For instance, during the construction of a dam on the Euphrates River (Gilli, 2015), a 100,000 m^3 chamber, the Crab Cave, was discovered 300 m below the dam site. It was filled with cement. Grout curtains were put in place, but significant leaks appeared up to 26 m^3/s, creating a vortex on the lake's surface.

The emptying of the lake made it possible to discover a second cave, the Petek Cave. It was filled with one million m^3 of cement and part of the left bank of the reservoir was covered with a 50 cm-thick curtain. This reduced the leaks that dropped down to 9 m^3/s.

The risk was underestimated because the limestone outcrops did not show evolved karst features very clearly. Former studies suggested that the underground caves had a hydrothermal origin. A similar problem occurred during the construction of the Atatürk Dam downstream of the Keban Dam where the discharge of hydrothermal springs increased gradually to over 10 m^3. A curtain wall was built that could not stop the leaks (Ertunç 1999). This shows the difficulty of understanding the behavior of karst systems in a country where the geological history is so complex.

7.6 Eustatism and Active Tectonic

In addition to normal processes, Turkish karsts are driven by three important phenomena: active tectonics, volcanism, and eustatism.

Since Turkey is surrounded by the Mediterranean, Aegean, and Black Sea, the karst system is driven by eustatism. Indeed, the karst domains widely extend along Turkey's seashore and submarine karst springs are observed in different sites of the Mediterranean Sea like in Ovacık where several outlets discharge 0.750 m^3/s (Elhatip 2003).

The relations between Turkish karsts and sea-level variations are poorly studied, but it is obvious that the effects of these variations on the karst features, and on the cave, genesis are important. The drop of sea level was 120 m every 100,000 years due to glaciations. However, during the Messinian salinity crisis (-5.9 Ma), the Mediterranean Sea was dry and the level dropped down to 2500 m (Roveri et al. 2014).

In similar zones, for example, in Southern France, karst aquifers and cave systems were intensely affected by this salinity crisis (Bini 1994; Audra et al, 2004; Mocochain et al 2006). Therefore, one can argue that in the Taurus, Thrace, and Black Sea mountains and in Western Anatolia, these

Fig. 7.13 Dumanli, one of world's biggest karst spring. It is now flooded by the water of the Oymapinar lake reservoir

sea-level drops caused the karst aquifers to migrate deeply which provoked the formation of cave systems deep below the current sea level. It confirmed that in Bodrum, where a warm karst spring is located on the small island of Kara Ada, there is a deep karst aquifer. Another example exists in Finike, where the Suluin cave (also called Incirli) was explored down to 120 m below the sea level. Other deep caves will probably be explored in the future.

The second important factor that drives kart evolution is tectonics. Turkey is located at the meeting point of several tectonic plates. The African plate moves toward the north and forms a subduction zone in the south of Turkey which caused the uplift of the Taurus range. The Arabian plate migrates to the north-west which causes Turkey to move toward west. An important shear zone is observed in its northern part, where the North Anatolian Fault separates the Anatolian plate from the Black Sea Mountains.

Uplift movements are a cause of deepening and rejuvenation of the karst systems. They also provoke the incision of deep valleys that cut and divide the karst aquifers. The presence of the obruks in the Konya Closed Basin and the large poljes of the Western Taurus Mountain are probably related to the quick uplift that favors vertical drainage by deep karst systems instead of horizontal drainage by surface hydrographic networks.

Another virtually unexplored research domain is volcanism and hypogene karsts. Explorations in the Ankarana (Madagascar) suggest that fluids associated with volcanism ($C0_2$, H_2O, H_2S) play an important role in the genesis of deep cave systems (Gilli 2019). Volcanism could, therefore, be an important cause of karst evolution in zones like the Anatolian plateau where the Mercies and the Hasan Day are high volcanoes. For instance, upwelling of CO_2 in relation with dormant volcanoes or deep-rooted igneous activity sources was advanced by Bayarı et al. (2009) to explain the formation of the obruks (Bayarı et al. 2017).

7.7 Conclusion

The complex relief of Turkey, the wide range of soluble rocks within the whole country, the active tectonics, the eustatism, the volcanism, and the climate variations explain the important diversity of Turkish karsts.

Thanks to the dynamism of Turkish speleologists, and the regular increase in university studies, the knowledge of Turkish karsts is actively progressing, but large research fields remain poorly explored, like the effects of the Messinian salinity crisis or the hypogene processes. However, due to its rapid economic development, Turkey is experiencing an increase in problems that concern land planning and water resources management. This augurs well for the growing development of applied karstology studies which will have to be based on fundamental research. The study of Turkish karsts should, therefore, progress in the coming years, and there is no doubt that Turkey will offer interesting discoveries in the large domain of speleogenesis and karst processes.

References

Altunel, E., & D'Andria F. (2019). Pamukkale Travertines: A Natural and Cultural Monument in the World Heritage List, in Kuzucuoglu C., Ciner A., Kazanci N., (eds.), Landscapes and landforms of Turkey, Eds. Springer, 219–232, https://doi.org/10.1007/978-3-030-03515-0

Altunel, E., & Hancock P. L. (1993). Morphological features and tectonic setting of Quaternary travertines at Pamukkale, Western Turkey. Geol J 28: 335–346

Altunel, E. (1994). Active tectonics and the evolution of Quaternary travertines at Pamukkale, Western Turkey. PhD thesis, University of Bristol (UK), 236 p

Arikan, Y. (1985). Geology and petroleum potential of the Tuz Gölü Basin (in Turkish). J MTA 85-2:17–38

Audra, P., Mocochain, L., Camus, H., Gilli, E., & Clauzon, G. (2004). The effect of the Messinian Deep Stage on karst development around the Mediterranean Sea. Examples from Southern France, Geodinamica Acta 17(6): 27-38.

Bakalowicz, M. (1970). Hydrodynamique karstique: étude du bassin d'alimentation de la Manavgat (Taurus occidental, Turquie). Thesis 3rd cycle, Paris

Bayari, C. S., Ozyurt, N. N., & Kilani, S. (2009b). Radiocarbon age distribution of groundwater in the Konya Closed Basin, central Anatolia, Turkey. Hydrogeology Journal, 17-2, 347–365

Bayarı, C. S., Pekkan, E., & Özyurt, N. N. (2009a). Obruks, as giant collapse dolines caused by hypogenic karstification in central Anatolia, Turkey: analysis of likely formation processes. Hydrogeology Journal. 17(2): 327–345

Bayarı, C. S, Özyurt, N., Törk, K., Avcı, P., Güner, I. N., & Pekkan, E. (2017). Geodynamic Control of Hypogene Karst Development in Central Anatolia, Turkey in A. Klimchouk et al. (eds.), Hypogene Karst Regions and Caves of the World, Cave and Karst Systems of the World, https://doi.org/10.1007/978-3-319-53348-3_27

Bini, A. (1994). Rapports entre la karstification peri méditerranéenne et la crise de salinité messinienne, l'exemple du karst lombard (Italie) Kars-tologia 23: 33–53

Burdon, D.J., & Safadi, C. (1963). Ras El-Ain: The Great Karst Spring of Mesopotamia: A Hydrogeological Study. Journal of Hydrology, 1: 58–95.

Canik, B., & Çörekçioğlu, I. (1986). The Formation of Sinkholes (obruk) be-tween Karapinar and Kızören - Konya Karst Water Resources (Proceedings of the Ankara - Antalya Symposium, July 1985) IAHS publ. 161: 193–205

Cater, J. M. L., Gillcrist, J. R. (1994). Karstic Reservoirs of the Mid-Cretaceous Mardin Group, SE Turkey: Tectonic and Eustatic Controls on their Genesis, Distribution and Preservation. Journal of Petroleum Geology, 17(3): 253–278.

Çemen, İ, Göncüoğlu, M. C., & Dirik, K. (1999). Structural evolution of the Tuzgölü basin in Central Anatolia, Turkey. J Geol 107: 693–706

Doğan, U., & Özel, S. (2005). Gypsum karst and its evolution east of Hafik (Sivas, Turkey). Geomorphology, 71(3-4): 373-388.

Doğan U., & Yeşilyurt, S. (2004). Gypsum karst south of Imranlı. Cave Karst Sci 31(1):7–14

Doğan, U., & Yılmaz, M. (2011). Natural and induced sinkholes of the Obruk Plateau and Karapınar-Hotamış Plain, Turkey. Journal of Asian Earth Sciences 40:496–508; DOI: https://doi.org/10.1016/j.jseaes.2010.09.014

Doğan, U., Koçyiğit, A., & Gökkaya, E. (2017). Development of the Kembos and Eynif structural poljes: Morphotectonic evolution of the Upper Manavgat River basin, central Taurides, Turkey. Geomorphology 278: 105–120

Doğan, U., & Yesilyurt, S. (2019). Gypsum Karst Landscape in the Sivas Basin in Kuzucuoglu C., Ciner A., Kazanci N., Landscapes and landforms of Turkey. Springer, 197–206, https://doi.org/10.1007/978-3-030-03515-0

Elhatip, H. (2003). The use of hydrochemical techniques to estimate the dis-charge of Ovacık submarine springs on the Mediterranean coast of Turkey. Environmental Geology 43:714–719

Erol, O. (1986). The relationships between the phases of the development of the Konya-Karapinar obruks and the Pleistocene Tuz Gölü and Konya pluvial lakes. In: Günay G, Johnson IE (eds) Karst water resources. International symposium (1985) IAHS-AISH publ. 161: 207–213

Ertunç, A. (1999). The geological problems of the large dams constructed on the Euphrates River (Turkey), Engineering Geology, 51: 167-182.

Frech, F. (1916). Geologie Kleinasiens in Bereich der Bagdadbahn [Geology of Asia Minor as regards the Baghdad Railway]. Zeitschrift Deutsch Ge-sellschaft Geowissenschaft 68:1–326

Gilli, E. (2015). Karstology. Karsts, caves and springs. CRC Press, Taylor & Francis Group, New York USA, 244 p

Gilli, E. (2019). The Ankarana Plateau in Madagascar. Tsingy, Caves, Volcanoes and Sapphires. Coll. Cave and Karst Systems of the World, Springer Nature edit, 162 p. https://doi.org/10.1007/978-3-319-99879-4

Glover, C., & Robertson, A. H. (2003). Origin of tufas (cool-water carbonate) and related plateaus in the Antalya areas, SW Turkey. Geol J 38:1–30

Kuzucuoglu, C., Ciner, A., & Kazanci, N. (2019). Landscapes and landforms of Turkey. Springer, 620 p. https://doi.org/10.1007/978-3-030-03515-0

Lu, Y., Qi, L., & Zhang, W. (2016). Discussion on the Process of Deep Karst and Hydrothermal Karst. Proceedings of Deepkarst 2016. Origins, Resources, and Management of Hypogene Karst April 11–14, 2016 Carlsbad, New Mexico, Usap 135–144

Mocochain, L., Clauzon, G., Bigot J-Y., & Brunet, Ph. (2006). Geodynamic evolution of the Peri-Mediterranean karst during the Messinian and the Pliocene: evidence from the Ardèche and Rhône Valley systems canyons, Southern France. Sedimentary Geology 188-189: 219-233

Nazik, L., Poyraz M., & Karabıyıkoğlu, M. (2019). Karstic Landscapes and Landforms in Turkey, in Kuzucuoğlu C., Çiner A., Kazancı N., Landscapes and landforms of Turkey, Springer, 181–196, https://doi.org/10.1007/978-3-030-03515-0

Nazik, L., & Tuncer, K. (2010). Regional features of Turkey karst morphology. Turk J Speleology Karst Cave Sci 1:7–19

Nicod, J. (2000). Ras-el-Aïn (Syrie N) et le problème des sources karstiques artésiennes aux marges ou à l'intérieur des domaines arides (Ras-el-Ain (N Syria) and the problem of the artesian karst springs at the border or in-side of the arid areas) Bulletin de l'Association de Géographes Fran-çais 77–2: 189–199

Öziş, S. (1992). Antalya travertines, hydrology and geochemistry. PhD the-sis, University of Çukurova, Adana, Turkey

Pekkan, E. (2004). Investigation of the hydrogeochemical processes that affects the formation of Obruks as karstic depressions located in the Konya closed basin (in Turkish). M.Sc. Thesis, Institute of Science, Hacettepe University Ankara, Turkey, 82 pp

Penck, W. (1918). Die tektonischen Grundzüge westkleinasiens. Engelhorn Nachf, Stuttgart, 136 pp

Roveri, M., Flecker, R., Krijgsman, W. J., Lofi J., Lugli, S., Manzi, V., Sierro, F. J., Bertini, A., Camerlenghi, A., De Lange, G., Govers, R., Hilgen, F. J., Hübscherk, Ch., Meijer, P. Th., & Stoica, M. (2014). The Messinian Salinity Crisis: Past and future of a great challenge for marine sciences. Marine Geology 352: 25–58 http://dx.doi.org/https://doi.org/10.1016/j.margeo.2014.02.002

Schildgen, T. F., Yıldırım C., Cosentino, D., & Strecker, M. R. (2014). Linking slab break-off, Hellenic trench retreat, and uplift of the Central and Eastern Anatolian plateaus. Earth Sci Rev 128:147–168

UN-ESCWA and BGR (2013) Jezira Tertiary Limestone Aquifer System. Inventory of Shared Water Resources in Western Asia. Beirut. Chap. 24. https://waterinventory.org/sites/waterinventory.org/files/chapters/Chapter-24-Jezira-Tertiary-Limestone-Aquifer-System-web.pdf. Accessed on 12/02/2020

Waltham, T. (2002). Gypsum karst near Sivas. Cave and karst science 29-1, 39–44

8 Karst Springs of Turkey: Hydrogeology of the Kirkgözler Karst Springs, Antalya

Gültekin Günay

Abstract

The karst springs in the study area constitute the most significant discharge agent of the large carbonate reservoirs. The largest of these springs are Kırkgözler and Düdenbaşı springs. Both of these springs discharge from the BA aquifer. Other major springs in the area discharge from the lower travertine plateau and the carbonates of the Antalya Nappes. The travertine springs are Duraliler, Arapsuyu, Mağara, Kemerağzı, and Iskele springs. The Hurma and Kapuz springs discharge from the Upper Antalya Nappes. There are also numerous small springs discharging at the coastal line. Kırkgözler karst springs control the discharge of carbonate rocks (with the age of the Mesozoic) of Beydağları Autochthon and Antalya Nappes. It is also the source of the Quaternary-aged Antalya travertine, and alluvial deposits lie in the plain.

8.1 Introduction

The study area is located at the Mediterranean coast of Turkey and is bounded by the Aksu River Basin in the east, the Beydağları Mountain in the north and west, and the Antalya Bay in the south. The major reservoir rocks of the springs of interest are the Mesozoic carbonates and the Antalya travertine. The region is also adjacent to the Manavgat Basin, which also includes the Dumanlı Karst Spring, one of the world's largest single-point discharged karst springs (Karanjac and Günay, 1980).

G. Günay (✉)
Hacettepe University, Yaşamkent Mah. Ataşehir Sitesi 7A- 5.Blok No:12, Çankaya, Ankara, Turkey
e-mail: ggunay@ttmail.com

8.2 Geologic Setting

The study area is located in a complex geological and structural environment of the Taurus Mountains, an extension of the Alpine orogenic belt into southwestern Turkey Brunn et al. (1971). The Western Taurus forms an arcuate belt divided into two limbs into both sides of the Antalya Bay. The Western Taurus consists of a central, relatively autochthonous unit, the Beydağları Autochthon. Two allochthonous ophiolite units, the Lycian (Elmalı) Nappes to the northwest and the Antalya Nappes to the east, lie on both sides of this unit (Günay and Bölükbaşı, 1981). The geologic map of the Kırkgözler zone and the Antalya travertine plateau is given in (Fig. 8.1).

8.3 Hydrogeology

8.3.1 Aquifers

At least three distinct groundwater-bearing bodies occur in the study area. In downward succession, they are the Mesozoic carbonate rocks of BA (Beydağları Autochthon) and the Antalya Nappes. The Quaternary-aged Antalya travertine and alluvial deposits from recent ages lie on the plains (Burger, 1990).

The carbonate rock units, outcropping in the north and northwest of the travertine deposits, exhibit an aquifer with large storage reservoirs. The carbonates of both the Beydağları Autochthon (BA) and the Antalya Nappes act as an aquifer, but the extend of carbonates of the Antalya Nappes is bounded by the impervious nappe units. These aquifers discharge through small springs, whereas the BA aquifer discharges through large ones at different elevations, such as the Kırkgözler and Düdenbaşı springs. The BA aquifer outcrops in high mountainous areas with an elevation range from 300 to 3000 m and bounded by the Lycian Nappes in the west, the Antalya Nappes in east and north, and the

G. Günay et al., *Caves and Karst of Turkey - Volume 2*, Cave and Karst Systems of the World,
https://doi.org/10.1007/978-3-030-95361-4_8

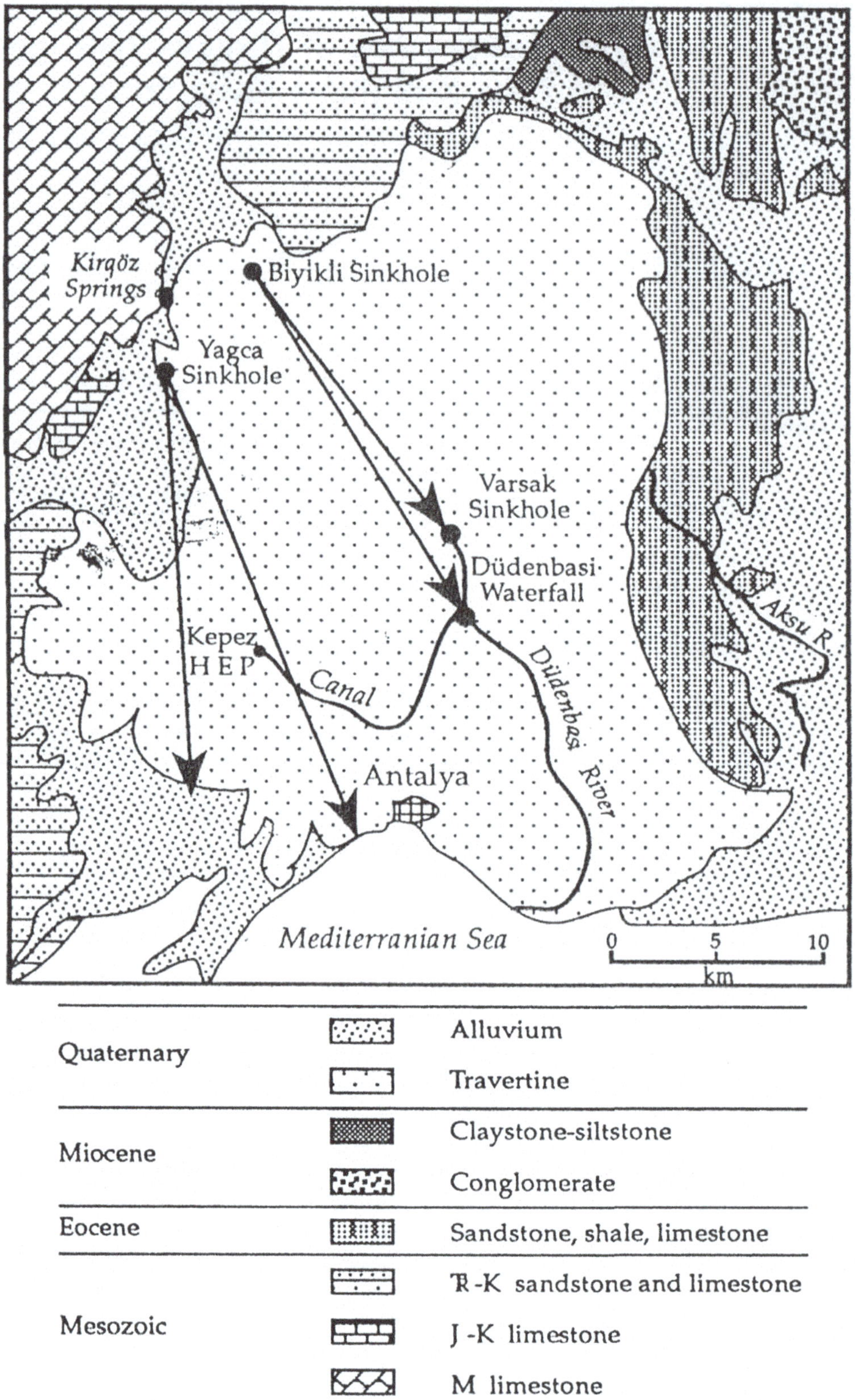

Fig. 8.1 Map showing geologic control on the flow of surface and groundwater in travertine terraces of Antalya (Back and Günay, 1992)

Miocene deposits in the south. The outcrop area of the BA aquifer is about 11,000 km^2. The recharge area of the aquifer extends in the SW-NE direction from the north of the Finike to the Eğridir Lake. The major natural discharge of this large area is the Kırkgözler Springs, whereas some small size discharges are observed at the Miocene boundary in the south, and the Lycian nappe boundary in the west.

The Antalya travertine is another significant water-bearing unit in the study area. The surface area of the travertine is about 630 km^2, and it acts as an unconfined aquifer.

8.3.2 Karst Springs

The karst springs in the study area constitute the most significant discharge agent of the large carbonate reservoirs. The largest of these springs are the Kırkgözler and Düdenbaşı springs. Both of these springs discharge from the BA aquifer (Eroskay and Günay, 1980). Other major springs in the area discharge from the lower travertine plateau and the carbonates of the Antalya Nappes (DSİ, 1985). The travertine springs are Duraliler, Arapsuyu, Mağara, Kemerağzı, and Iskele springs. The Hurma and Kapuz springs discharge from the Upper Antalya Nappes. There are also numerous small springs discharging at the coastal line (Figs. 8.1 and 8.2).

8.3.2.1 Kirkgözler Springs

The springs discharge from the eastern foothills of the Katran (Beydağları) Mountain along the zone of 1 km length at the elevation of 300 m. The discharge occurs from many outlets, whereas the name of the Kırkgözler Springs represents the entire group. Some of the larger outlets have special names such as Pınarbaşı, Karagöz, Incirligöz, Boynuzlugöz, and Kocain cave. All the outlets are discharging the same reservoir of the BA aquifer through well-developed conduits and fractures. The spring zone is located in the contact zone of the BA limestone and the travertine. However, the occurrence of the springs is controlled by the subsurface extension of the impermeable the Antalya Nappes units, mainly the melange units of the Middle Nappe (Figs. 8.2, 8.3 and 8.4).

8.3.2.2 Düdenbaşi Spring

The Düdenbaşı spring discharges at 80 m, 9 km northeast of Antalya. The outlet of the spring is located on the lower travertine plateau, however, discharges the BA aquifer. The water rises from the bottom of a collapse with a diameter of 400 m. The groundwater of the BA aquifer is conveyed via a conduit directed to the NW, toward the BA limestone units. The entrance of the conduit starts few meters below the water level. The spring water creates the Düden River flowing south to the Düden regulator and then to the Mediterranean Sea along a route of 15 km. Part of the river water is used for irrigation by the local people. The cross-section of the Düdenbaşı resurgence is given in (Fig. 8.5).

The groundwater discharge through the springs is one of the major components of the water budget. Many springs are discharging in the study area, and only a small portion of their discharges has been measured. The discharge rate of the coastal, submarine, and small mountainous springs has never been measured. The major springs and their long-term average discharge rates are given in Table 8.1.

This high value of the groundwater discharge takes place in all the aquifer materials including the BA aquifer, the Travertine Aquifer, and the Antalya Nappe Aquifer. The discharge of the Kırkgözler Springs and Düdenbaşı occupies 73% of the total measured flows in the area (Günay et al., 1995; Tezcan et al., 1997).

8.4 Hydrochemistry

During the study period, 14 springs, 2 surface water, and 2 wells have been sampled continuously between April 1995 and August 1996 every month for the hydro-chemical analyses. Five of the springs are different outlets of the Kırkgözler

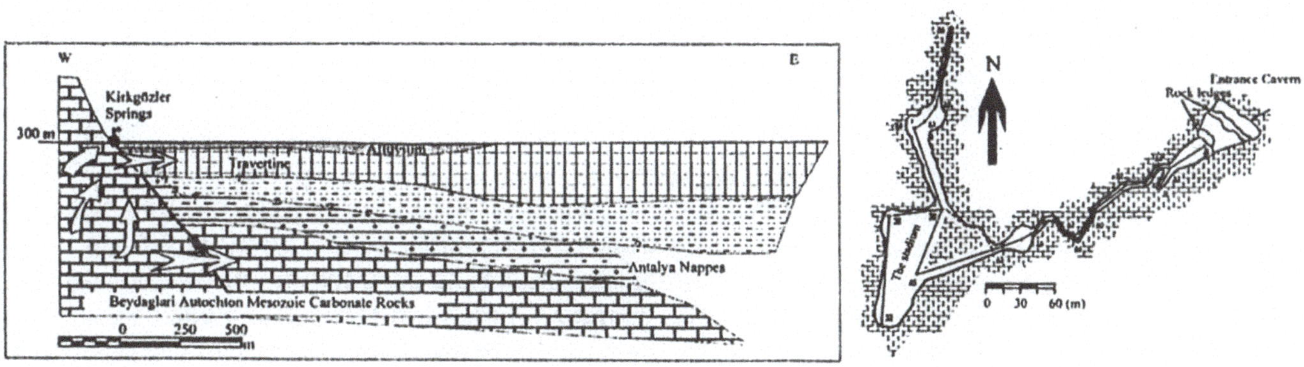

Fig. 8.2 The geological cross-section for the occurrence of the Kırkgözler springs and the sketch map of the Kırkgözler cave system (after Kincaid 1996)

Fig. 8.3 Kirkgözler karst springs main outlet

Table 8.1 The long-term average discharge rates of major springs

Spring	Average Flow (m^3/s)	Aquifer
Düdenbaşı spring	18.3	Beydağları Autochthon
Kırkgözler springs total	15.3	Beydağları Autochthon
Coastal springs	5.0	Antalya travertine
Duraliler	1.8	Antalya travertine
Mağara	1.5	Antalya travertine
Kemerağzı	1.4	Antalya travertine
Arapsuyu total	0.9	Antalya travertine
Kapuz total	0.8	Upper Antalya nap
Hurma	0.6	Upper Antalya nap
Total	45.6	

Springs group. The sampling period covers two high water seasons (April–Jun 95/96) and two low water conditions (Oct 95/Sep 96). The temperature, and the specific electrical conductivity, pH, CO_2, and DO were measured in the field, whereas the HCO_3—was analyzed the same day of the sampling or the day after. The cations (Ca^{+2}, Mg^{+2}, Na^+, K^+, Sr^{+2}) and other anions (Cl^-, SO_4^{-2}, NO_3^-) were analyzed at the hydrochemistry laboratory of the International Research and Application Center for Karst Water Resources of Hacettepe University. In low water conditions, October 1995 and 1996, the heavy metals were also analyzed to check the composition in the reservoir system. The cations were preserved by keeping pH below 2, and the analyses were completed almost one week after the sampling. All the analyses and sampling

Fig. 8.4 Kirkgözler karst springs several outlets (Günay and Yayan, 1979)

were carried out by the same people. Unbalanced analyses were repeated until the electrical neutrality conditions were obtained. The electrical neutrality of the analyses was below 5%. The saturation indices and the total dissolved solids (TDS) content were calculated by the computer program WATEQ (Truesdell and Jones 1973).

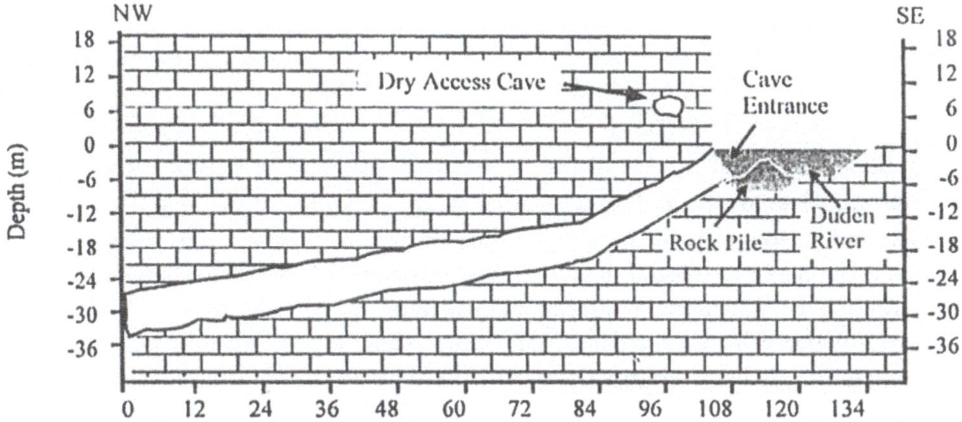

Fig. 8.5 Sketch map of the Düdenbaşi resurgence and cave system (after Kincaid 1996)

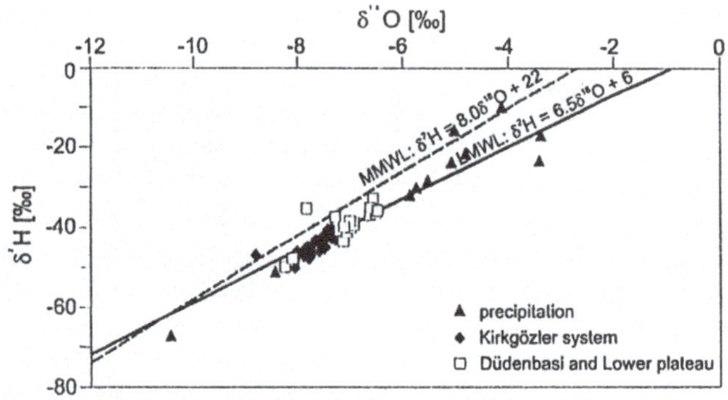

Fig. 8.6 Stable isotope composition of the precipitation and all springs sampled during 1995 and 1996 (mmwl = mediterranean meteoric water line (Yurtsever 1979) and LMWL = local meteoric water line (Nativ et al. 1997)

8.5 Natural and Artificial Tracing

For the natural labeling, a set of environmental tracers (18O, 2H, 3H, 13C, 14C, CFCs, He) was used (Nativ et al. 1999). The sampling took place parallel to the sampling for the hydro-chemical characterization. However, additional monthly precipitation samples from the Antalya stations on the coast and Döşemealtı on the upper plateau were also taken. The samples were analyzed in the usual way. A detailed description can be found in Nativ et al. (1999). The mean values and range of isotopes were investigated in the groundwater, and the precipitation was depicted. The stable isotope relation of the precipitation and all springs during, including their relation to the Mediterranean (Yurtsever 1978) and the Local Meteoric Water Line (Nativ et al. 1999), were sampled during 1995 and 1996. In addition, the seasonal variation in 18O and 3H in the precipitation, and three springs selected according to their position, is presented in (Fig. 8.6).

8.6 Conclusions

The hydrogeological structure of the Kırkgözler springs was characterized by a combined application of various investigation methods to provide descriptive information for reasonable management of the Beydağları carbonate and Antalya travertine aquifer system. Depending on the huge recharge area, the available water quantity seems to be sufficient for the future. The seawater intrusion has not been recorded up until now, although several coastal springs and exploitation well fields situated near the coast discharge the travertine aquifer. The quality parameters have not exceeded the permissible standard values so far, but intensive agriculture and the lack of adequate sewage systems could lead to a significant degradation in the long term. The existence of a conduit flow (characterized by high now velocities of more than 200 m/h), although their volume is smaller than that of the less permeable parts, exhibits a special risk factor in case of a concentrated pollution input, a conduit flow characterized by high now velocities of more than 200 m/h still exists. To preserve the high water quality standard, all domestic, industrial, and agricultural effluents should be controlled and disposed of, in addition, treatment plants should be built.

References

Back, W., & Günay, G. (1992). Tectonic Influences on groundwater flow systems in karst of the southwest Taurus mountains, Turkey. in W. Back, J. S. Herman & H. Paloc (Eds.), Hydrogeology of Selected Karst Regions, International Contributions to Hydrogeology, Vol. 13, 263–272 p., Verlag Heinz Heise, Hannover, FRG.

Brunn, J. H., Dumont, J. F., Graciansky, P. C., Gutnic, M., Juteau, T., Marcaux, J., Monod, O., & Poisson, A, (1971). Outline of the geology of the western Taurids. In: Campell, A. S. (ed). Geology and history of Turkey, Petrol. Explor. Soc., Libya, 225–252.

Burger, D. (1990). The travertine complex of Antalya, SW Turkey, Z. Geomorph. N. E., 77, pp. 25–46.

DSİ. (1985). Antalya Kırkgöz kaynakları ve traverten platosu karst hidrojeolojisi etüd raporu, DSİ Jeoteknik Hizmetler ve Yeraltısuları Dairesi Başkanlığı, Ankara, 131 sf. (in Turkish).

Eroskay, S. O., & Günay, G., (1980). Tecto-genetic classification and hydrogeological properties of the karst regions in Turkey. in G. Günay (Ed.), Karst Hydrogeology Proceedings: October 1979, Oymapinar-Antalya, Turkey, UNDP Project TUR/77/015, p. 1–41.

Günay, G., Tezcan, L., Ekmekçi, M., & Atilla, Ö., (1995). Present state and future trends of karst groundwater pollution in Antalya travertine plateau, Turkey. Europe. Commission, Report EUR 16547, 305–324.

Günay, G. & Yayan, T. (1979). Antalya - Kırkgöz kaynakları hidrojeoloji incelemesi. 1.Ulusal Hidrojeoloji Semineri, DSİ Oymapınar Barajı, Antalya. DSİ-UNDP Projesi TUR / 77/ 015 Project Studies. DSİ Groundwater Dept. Yücetepe, Ankara (in Turkish).

Günay, Y., & Bölükbaşı, S., (1981). Antalya-Elmalı-Korkuteli-Bucak arasındaki Beydağlarının jeolojisi ve petrol olanakları, TPAO Rapor No: 1566, Ankara (in Turkish).

Karanjac, J., & Günay, G. (1980). Dumanlı spring Turkey - the largest karstic spring in the world, Journal of Hydrology, 45, p. 219–231.

Kincaid, T. R. (1996). Project Karstdive 95. unpublished, 66 p., University Wyoming, USA.

Nativ, R., Günay, G., Tezcan, L., Hötzl, H., Reichert, B. & Solomon, K., (1997). Separation of groundwater-flow components in a karstified aquifer using environmental tracers. in: A. Kranjc (Ed.), Tracer Hydrology. 97, 269–272, CRC Press.

Nativ, R., Günay, G., Hötzl, H., Reichert, B., Solomon, D.K. & Tezcan, L., (1999). Separation of groundwater-flow components in a karstified aquifer using environmental tracers. Appl Geochem 14:1001–1014.

Tezcan, L., Günay, G., Hötzl, H., Reichert, B. & Solomon, K., (1997). Hydrogeology of the Kırkgözler Springs, Antalya, Turkey. International Conference on Water Problems in the Mediterranean Countries. 17–21, November 1997, Near East Technical University, Nicosa, North Cyprus.

Truesdell, A., & Jones, B., (1974). WATEQ: A computer program for calculating chemical equilibria of natural waters. J. Res. U.S. Geol. Surv. 2:233–248.

Yurtsever, Y., (1978). Environmental isotopes as a tool in hydrogeological investigation of southern karst region of Turkey. Proceedings of a International Seminar on Karst Hydrogeology, Antalya, Turkey.

The Karst Springs of Antalya, Turkey

9

Gültekin Günay

Abstract

The Olukköprü springs, the Böğrümköprü, Alabalık pool, and Oğlanuçtuğu springs in south are discharges from the Köprüçay canyon. The Olukköprü springs discharge water to the Köprüçay River. The largest contribution to the Köprüçay River flow is from The Olukköprü springs. The Olukköprü springs discharge water on both sides of the Köprüçay River that flows into the sea. It drains an area bounded by the Beyşehir and Eğridir lakes in the north, the Aksu and Manavgat in east.

9.1 Introduction

The Olukköprü springs are known that springs are given special importance is attached to the springs regarding the water resources projects in the adjacent basin. The Manavgat basin is the adjacent basin at the east part of Köprüçay basin and includes the Dumanlı Karst Spring, one of the world's largest single-point discharged karst springs (Günay and Yayan, 1979; Karanjac and Günay, 1980). The stratigraphic units in the region can be divided into allochthonous units, in terms of their lithological and structural features. The Paleozoic, Mesozoic, and Cenozoic rocks prevail in the region (Akay and Uysal, 1985). The Cenozoic rocks are the oldest. Köprüçay Basin and comprise gradually alternating conglomerates Köprüçay. Beşkonak formation is impermeable.

9.2 The Oluköprü Springs

Along the right bank of the Köprüçay River, many springs occur above the river level. The springs discharge through the bedding planes. The Olukköprü another spring around it is discharge the same aquifer. All spring discharges occur along the Köprüçay River, Kırkgeçit Stream, and Oğlanuçtuğu fault (Figs. 9.1, 9.2 and 9.3).

9.3 Spring Discharge Rates

In a flow the Olukköprü springs are located Beşkonak that there is a seasons precise analysis of the springs (Değirmenci and Günay, 1990). The Köprüçay flows, Olukköprü. Böğrümköprü, Alabalık pool, and Oğlanuçtuğu springs discharges are calculated: the 31 m^3/s for Oğlanuçtuğu springs; 2.5 m^3/s for Böğrümköprü springs; and 6 m^3/s for both Alabalık pool and Oğlanuçtuğu springs, respectively (Değirmenci 1989).

9.4 Recharge Areas of the Springs

To determine the recharge areas of the springs were sub-basins of the Köprüçay and the rainfall—evaporation—surface runoff was made. Olukköprü and aquifers are based mainly on the Technical Report Of the project "The Regime of the Olukköprü and Kocadere Springs of the Köprüçay basin" (Karanjac and Altuğ 1976) The Olukköprü springs discharge water on both sides of the Köprüçay River that flows into the sea. It drains an area bounded by the Beyşehir and Eğridir lakes in the north, the Aksu Basin west, Manavgat east (Değirmenci and Günay, 1993). Narrow

G. Günay (✉)
Faculty of Engineering, Hacettepe University, Beytepe, Ankara, Turkey
e-mail: ggunay@ttmail.com

G. Günay et al., *Caves and Karst of Turkey - Volume 2*, Cave and Karst Systems of the World,
https://doi.org/10.1007/978-3-030-95361-4_9

Fig. 9.1 The one of outlet of Köprüçay–Olukköprü karst springs. The outlet is above about 1.5 m from river the river bed (view is from south to north)

Fig. 9.2 The one of the main outlets of Köprüçay–Olukköprü karst springs in Beşkonak village

Köprüçay River gorges with the emerging springs. Springs emerge in a large area, including the widened valley downstream of the gorge, and the right bank tributary Koca Creek, with several dozen small, medium, and large springs. The entire spring area near the Beşkonak village, some of which are a few meters above the river level, is a national park and recreation area.

9.5 Allochthonous Units

Units of Antalya nappes occur between the Köprüçay Basin and the Aksu Basin to the west of the region. Antalya nappes cover large areas of the Aksu and Köprüçay Basins, from Eğridir Lake in the north to the Serik-Aspendos area in the south. Due to their lithological and structural properties, permeable and impermeable nappe units in this area are of great importance from the geotechnical and hydrogeological points of view.

In the Aksu and Köprüçay Basins, Antalya nappes are in contact with the underlying Jurassic-Cretaceous limestones. In the study area, the stratigraphic sequence from the bottom to the top is as follows: Ophiolitic unit/Ispartaçay unit (Tr-K), Dulup unit/Tahtalıdağ Group (P-K) (Şenel, 1997).

9.6 Conclusion

The discharge measurements at the Bulasan and Beşkonak stream gauging stations, 1963–1971 period Köprüçay River flows come through the Olukköprü springs.

If waters of the Beyşehir and Eğridir waters feed the aquifer of the Olukköprü Springs, then they have a long route of about 50 km or more.

Fig. 9.3 The one of outlet of Köprüçay–Olukköprü karst springs. The outlet is above about 1 m from river the river bed about (view is from southwest to northeast)

References

Akay, E. & Uysal, Ş. (1985) Orta Torosların batısındaki (Antalya) Neojen çökellerin stratigrafisi sedimantalojisi ve yapısal jeolojisi. MTA Rapor No: 7799 (in Turkish).

Değirmenci, M. (1989). Köprüçay ve dolayının (Antalya) karst hidrojeoloji incelemesi. Doktora tezi. Hacettepe Üniversitesi Fen Bilimleri Enstitüsü, Beytepe, Ankara, 398 sf. (in Turkish).

Değirmenci, M. & Günay, G. (1990). Analysis of hydrologic relations between Eğirdir Beyşehir-Suğla Lakes systems and adjacent basins by means of remote sensing techniques, (South Turkey). Environmental Geology and Water Sciences ,19, 1, 41–45.

Değirmenci, M. & Günay, G. (1993). Origin and catchment area of Olukköprü karst springs. in: G. Günay, A. I. Johnson and W. Back (Eds.). Hydrogeological processes in karst terrains, Proceedings of the Antalya Symposium and Field Seminar, IAHS Publication No: 207, 97–106.

Günay, G. & Yayan, T. (1979). Antalya - Kırkgöz kaynakları hidrojeoloji incelemesi. 1. Ulusal Hidrojeoloji Semineri, DSİ Oymapınar Barajı, Antalya. DSİ-UNDP Projesi TUR / 77/ 015 Project Studies. DSİ Groundwater Dept. Yücetepe, Ankara (in Turkish).

Karanjac, J. & Altuğ, A. (1976). Regime of Olukköprü and Kocadere springs of Köprüçay basin based on hydrograph analysis. Technical Report of UNDP-DSİ project, DSİ, Ankara.

Karanjac, J., and Günay, G., 1980, Dumanlı spring Turkey - the largest karstic spring in the world. Journal of Hydrology, 45, 219–231.

Şenel, M., 1997, Isparta ve Antalya jeolojik paftaları (1:100000 ve 1:250000 ölçekli), MTA, Ankara (in Turkish).

Karst Hydrogeology of Pamukkale Thermal Springs, Denizli, Turkey

10

Gültekin Günay

Abstract

Pamukkale thermal waters are generally composed of karstic, highly fractured, and cracked Paleozoic-aged marbles as the main aquifer. The flow rate of the springs is around 385 l/s. Environmental isotope assessments have shown that hot waters have a long underground water circulation system (approximately 20–30 years).

10.1 Introduction

As it is declared by UNESCO, the Cotton Castle “Pamukkale” has a historical value besides geological importance ((Figs. 10.1, 10.2 and 10.3). The thermal spa called Hierapolis was established in the area by Pergamon at the end of the second century. The area with ruins and Pamukkale hydrothermal travertine terrace site is 20 km away from Denizli Province in the central Aegean region.

10.2 Hydrogeological Properties

Paleozoic-aged marbles forming the main aquifer show very fractured and karstic features. The marbles are bounded by Pliocene-aged thick, semi-permeable, and impermeable units. A thick covering formation spa catchment basin extending along Yenice Horst is covered with Pliocene-aged limestone (Şimşek, 1990). The annual average areal precipitation is 583 mm calculated by the isohyetal method. There are four mainspring water sources on the travertine. All the source exits are along the main fault line in the direction of NW-SE.

10.3 Hydrogeochemical Aspects

A weekly observation, measurement, and sampling program were conducted to explain the feeding and flow regime of hot springs. The information helped to establish a conceptual model for the regional hydrodynamic structure (Günay et al., 1997). However, the precipitation kinetics of travertine was also investigated. Travertine accumulation is greatest at the edges of the terraces, where the outflow of carbon dioxide gas increases with a turbulent flow. Pamukkale thermal springs (UKAM, 1994).

10.4 Pollution and Protection Studies

The color of the white travertine started to change with the arrival of hotels in the travertine areas. The hotels take hot water directly to the swimming pool before they release it onto the travertine. The water in the pools decreases its travertines depositing capacity, and also people in the swimming pool leave some organic materials which cause rapid growth of algae (UKAM, 1995). The lack of a sewage system, and the presence of septic tanks for every hotel, is the other major source of pollution. The septic tanks are dug in the travertines, and also, they are lined with cement.

10.5 Conclusion

The several types of the structure of the area were explained: Touristic activities (hotels, shops, etc.) will be reduced and removed in all the travertine areas. Stepping on the travertines and the traffic on the site will be strictly prohibited (Figs. 10.3 and 10.4). Septic tanks will be removed,

G. Günay (✉)
Hacettepe University, Ankara, Turkey
e-mail: ggunay@ttmail.com

G. Günay et al., *Caves and Karst of Turkey - Volume 2*, Cave and Karst Systems of the World,
https://doi.org/10.1007/978-3-030-95361-4_10

Fig. 10.1 Main big carbonate terraces of Pamukklae travertines. The direction of photo is from west to east

Fig. 10.2 Cotton white travertine rimstones and artificial concrete pools which was built on the removed asphalt road in 1995

Fig. 10.3 The wooden walkway was built to protect travertine from stepping of visitors in 1997

Fig. 10.4 Artificial concrete pools for visitors on the old asphalt road removed 1995

instead, mobile toilets will be provided. A special security system was established. The water distribution system was established, and technical staff and also hydrogeologists were assigned. Entrance to the travertine and walking on the white travertine area is banned.

References

Günay, G., Şimşek, Ş., Keloğlu, N., Ekmekçi, M., Elhatip, H., Yeşertener, C., Dilsiz, C. & Çetiner, Z. (1997). Karst hydrogeology and environmental impacts of Denizli-Pamukkale springs. in G. Günay & A.I Johnson (Eds.). Karst Waters and Environmental Impacts, Balkema-Rotterdam, ISBN 90-5410-8584.

Şimşek, Ş. (1990) Karstic hot water aquifers in Turkey. in G. Günay, A. I. Johnson & W. Back. Proc. of the Int. Symp. On Hydrogeological Processes in Karst Terranes, Antalya-Turkey, 173–184 p., IAHS Publ. No. 207, UK.

UKAM. (1994). Project Phase-I Report on Conservation and Development of Pamukkale Travertines. Hacettepe University, Int. Res. Centre For Karst Water Res. - Ministry of Culture of Culture. Techical Report, 132 p. (in Turkish, unpublished).

UKAM. (1995). Project Phase-II Report on Conservation and Development of Pamukkale Travertines, Hacettepe University, Int. Res. Centre For Karst Water Res. - Ministry of Culture of Culture. Techical Report, 83 p. (in Turkish, unpublished).

Beyazsu and Karasu Karst Springs Mardin-Nusaybin Area (SE Turkey)

11

Gültekin Günay

Abstract

The carbonate rock masses that form the deep aquifer system in the plain correspond to the lower part of the upper autochthonous succession. The Eocene Midyat limestone is highly karstified. Karstic springs discharging from the Midyat formation are located mainly at the eastern part where this formation outcrops over large areas. The hydrogeological properties of the lithological units are described mainly based on the data obtained from the shallow and deep wells.

11.1 Introduction

The Kızıltepe Plain, having an area of about 600 km^2 at the Turkish-Syrian border in Southern Turkey, is of great importance for the socio-economic development of the region. Although the topography and soil structure favor agriculture, water shortage is the major factor hindering the development of the plain. The multi-aquifer system in the plain consists of three individual aquifers whose water might be available for irrigation and domestic use. The upper aquifer is made of unconsolidated clastic and basal conglomerate. The static water level at this aquifer ranges between 5 and 30 m from the surface. This aquifer is separated from the underlying limestone aquifer by an impermeable unit made up of marl and clay. The water level of the intermediate limestone aquifer is lower than 50 m below the surface. The secondary porosity of this aquifer is relatively high. Some solution cavities were also encountered during drilling. The lower aquifer, which is located at greater depth, is made of fractured and karstified limestone and dolomite from the Upper Cretaceous age. The intermediate and lower aquifers are separated by chalky marl and marl.

11.2 Geology of the Deep Aquifer Systems

According to Yalçın (1976), the carbonate rock masses that form the deep aquifer system in the plain correspond to the lower part of the upper autochthonous succession (Fig. 11.1). Yılmaz (1993) suggests that in the areas of the foreland that are not covered by the Upper Cretaceous ophiolite, marine sedimentation of mainly neritic carbonate rocks laterally grades into shale (The Germav formation). The Germav formation is overlain by a new level of clastic rocks across an unconformity (the Gercüş formation), which is gradually replaced upward by a thick carbonate succession of the Midyat group. This group is from the Eocene age and forms the upper part of the deep carbonate aquifer system. The Eocene carbonate rock series, named the Midyat formation, presents a succession of clay and dolomitic limestone in the lower part, thin-bedded marly limestone with some chert layers in the middle and massive limestone in the upper part. The thickness of the Midyat group reaches up to 1000 m. A flysch succession from the Miocene age, as described in the previous section, overlies the Midyat group conformably.

On the other hand, the lithology that is underlying the Midyat group belongs to the lower autochthonous succession which comprises Paleozoic and Mesozoic deposits ranging from the Precambrian to the Upper Cretaceous. From the Devonian to Cretaceous, the clastic deposits are replaced by neritic carbonate rocks. These carbonate rocks form the lower deep aquifer in the Kızıltepe Plain. The succession in Mesozoic starts with Triassic marly limestone which displays lateral changes (Yılmaz 1993). A moderately thick (500 m) alternating beds of shallow-water limestone and dolomite named as Cudi formation overlie the Triassic units. The Cudi carbonate rock series is from the Jurassic age and is overlain by a thick succession that comprises various lithologies of both clastic and carbonate rock series.

G. Günay (✉)
Hacettepe University, Yaşamkent Mah. Ataşehir Sitesi 7A-5. Blok No: 12, Çankaya, Ankara, Turkey
e-mail: ggunay@ttmail.com

G. Günay et al., *Caves and Karst of Turkey - Volume 2*, Cave and Karst Systems of the World,
https://doi.org/10.1007/978-3-030-95361-4_11

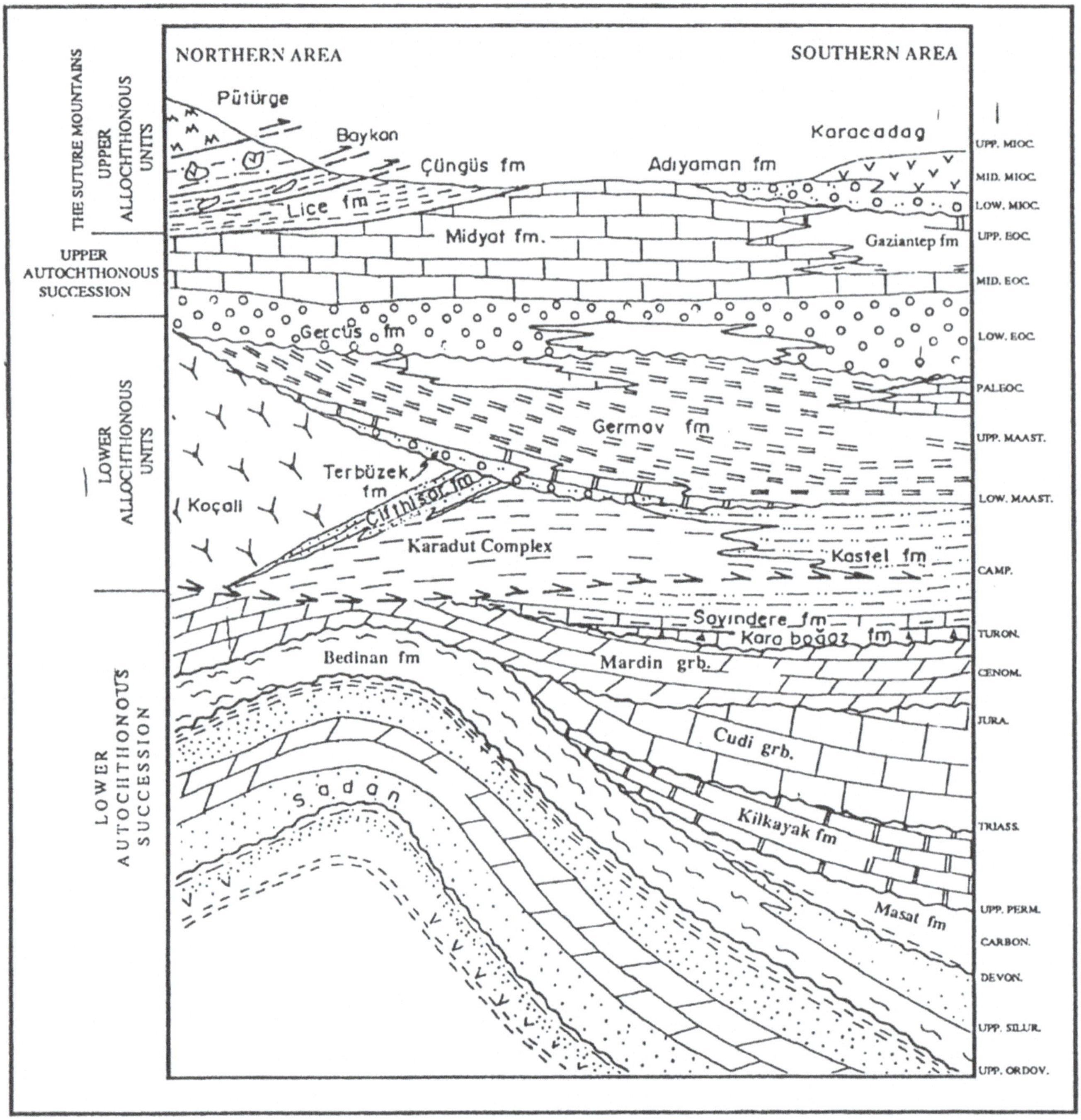

Fig. 11.1 Generalized geological section of the northwestern portion of the Southeastern Anatolia. Modified From Yalçın (1976); Eroskay and Günay (1980)

11.3 Hydrogeological Setting

The Southeastern Anatolia geographical region is characterized by a semi-arid to the arid climate. The mean annual precipitation ranges between 550 and 850 mm. This figure increases toward the north where the watershed of the aquifer system extends over large areas. For instance, the annual precipitation exceeds 1200 mm in Gercüş and Haberli meteorological stations, which are located in the northern part of the area for several years. Recharge to the upper-shallow aquifer system occurs through the taluses and conglomerates outcropping in the north of the plain, and partly through the clayey sand. The recharge is limited in the central parts because the conglomerate and talus are overlain by sandy clay, and clay and marl in the east and south.

Therefore, the most upper alluvial part contains shallow groundwater that is available through shallow dug wells.

The Midyat limestone outcrops at very limited areas (80 km^2) in the west whereas it has a much larger areal extent in the east. Karstic springs discharging from the Midyat formation are located mainly in the eastern part, where this formation outcrops over large areas. The discharge rate of the Beyazsu (Figs. 11.2 and 11.3) and Karasu karst springs were measured in February 1985 as 4.25 m^3/sec and 4 m^3/sec respectively (DSİ, 1987). The hydrogeological properties of the lithological units are described mainly based on the data obtained from the shallow and deep wells. The Quaternary talus comprising coarse blocks and gravel mainly form the unsaturated zone whose thickness varies from 1 to 10 m. The underlying alluvium is made up of day, sand, and gravel and contains a small quantity of water, especially, where sand and gravel are dominant. According to the pumping tests carried out by DSİ (1970), the transmissivity coefficient varies between 50 and 2000 m^3/day/m, depending on the thickness of the saturated zone which ranges between 10 and 80 m.

All other Miocene-Pliocene lithologies of flysch character, such as marl, clay, clayey sand, marly limestone, are poor of water. The hydrogeological properties of the deep carbonate aquifers differ significantly since they are highly karstified. The data available are obtained mainly from oil drilling and fieldwork. The limited data may suggest that the karstification of this limestone has developed in the past geologic times before the Miocene deposition, and therefore, it is most likely of paleokarstic character.

The Eocene Midyat limestone is highly karstified. However, it does contain water only when the recharge conditions favor it. The infiltration rate through the Midyat limestone is calculated as 80%, which provides an important quantity of water to the groundwater reservoir. All these figures suggest that there might be a significant groundwater potential in the deep Midyat limestone aquifer. The oil drillings that penetrated the lower deep Cretaceous limestone

Fig. 11.2 Beyazsu Springs discharging from Midyat limestone

Fig. 11.3 One of the outlet of Beyazsu Springs and Midyat limestone behind the scene

aquifer are again karstified. This is a paleokarstic aquifer and bears water in its openings. The examined oil drillings showed that there was no circulation at 600 m below the surface, and no return took place at greater depths down to 4500 m. geologic section (Fig. 11.1) and Beyazsu springs (Figs. 11.2, 11.3, 11.4 and 11.5).

11.4 Water Quality

The hydro-chemical analyses of samples collected from wells that penetrated the upper-shallow aquifer of the Miocene-Pliocene sediments revealed that the water quality is closely related to the lithology. In wells, receiving water from clastic sediments, the water is of lower hardness (15–20 °F), whereas its hardness reaches up to 75 °F in limestone. The discharging water is from a typical calcium-bicarbonate class with a hardness of about 33 °F. Toward south and southeast, this aquifer becomes deeper and the water chemistry differs from the upper levels. The total dissolved solid content does not exceed 2500 mg/L even in the deepest horizons (4000 m).

11.5 Deep Aquifer System

There are two ways of exploiting the deep Midyat limestone aquifer. The easier way, which is presently practiced, is to drill relatively shallow wells at localities close to the recharge area, where the clastic overburden is thin. This type of favorable area extends mainly in the eastern and southeastern parts of the plain. The static water level in this area ranges from 120 m near the karst springs down to 175 m.

11.6 Conclusions

However, various difficulties, inherent to the limestone aquifers, deterred the hydrogeologists from drilling deeper wells. There are two deep limestone aquifers, which are separated by impervious clastic sediments. The upper deep limestone aquifer, named the Eocene Midyat limestone aquifer, extends down to about 1000 m, whereas the lower deep aquifer, the Jurassic-Cretaceous aquifer, reaches about 2500 m below the surface. Due to the great depth and the

Fig. 11.4 Groundwater discharges from some outlets of Beyazsu springs are collected in the canal

limited recharge of the latter, it is not examined in detail in this study.

The promising deep aquifer (Midyat karstic aquifer) is worth exploring if the great need for water is considered. Although this aquifer has a high permeability, its outcrop is limited in the western part of the plain, and therefore, it is not recharged adequately to be exploited through deep wells. Large outcrops in the northeastern part provide a good recharge to the aquifer because the annual precipitation is as high as 1000 mm. The hydro-chemical analyses obtained both from relatively deep wells and from oil drillings revealed that the water quality is good enough to be used for domestic purposes, and it suggests that this deep aquifer may be replenished and bearing recent water. This should be verified by isotope analyses. Considering the geological setting, it is more convenient to exploit this aquifer in the east-southeastern part of the plain. Two huge karstic springs discharge from the Midyat karstic aquifer. The waters of the Beyazsu Springs are newly distributed to Mardin for the city supply. The existence of these springs at high altitudes

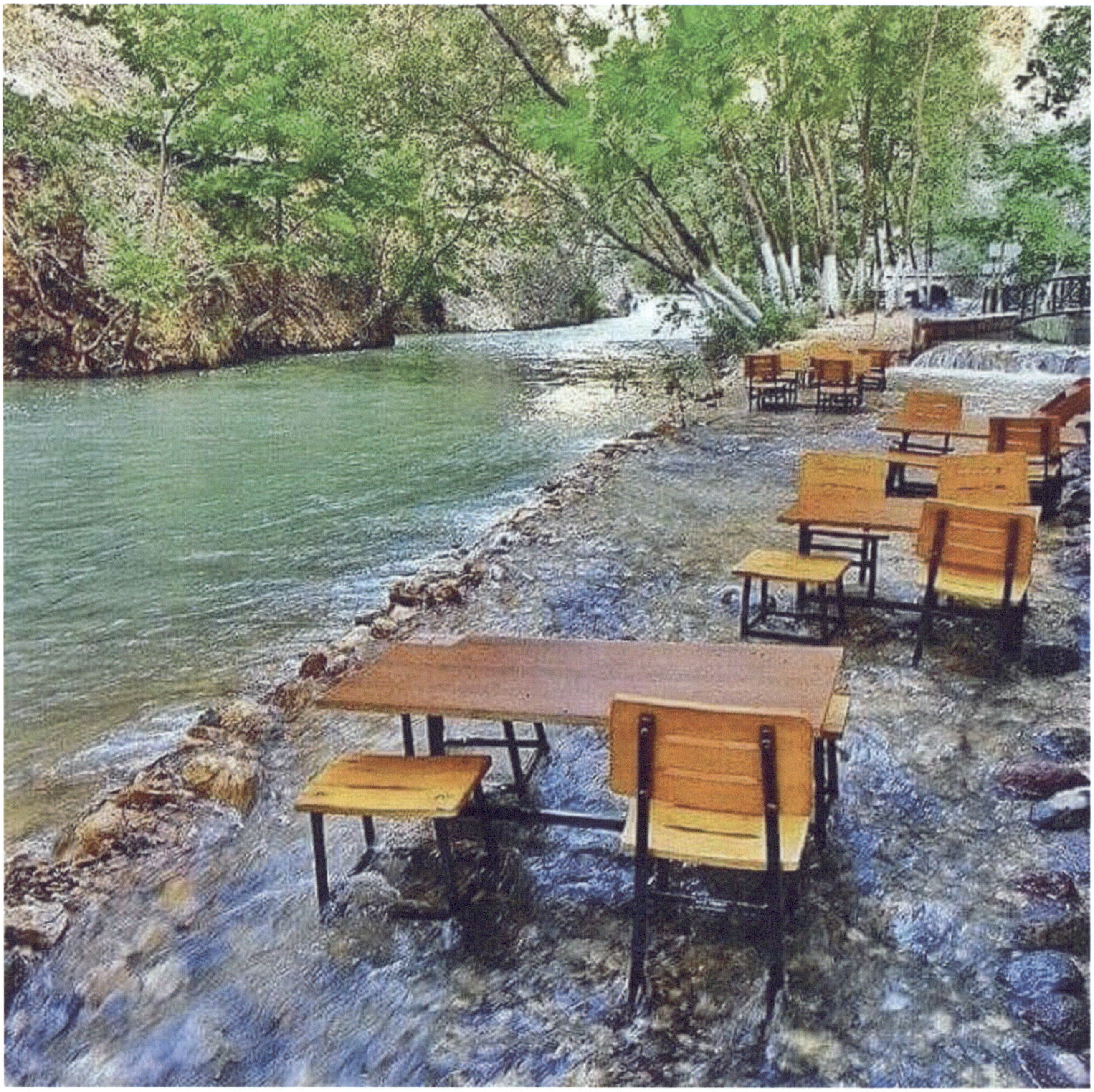

Fig. 11.5 Groundwater discharging from all outlets of Beyazsu springs are flowing in the river bed

(600–650 m) may reflect that the karstification base does not extend down to great depths. But, from the oil drillings, in which circulation of mud was lost when the drilling proceeded within this aquifer, it is apparent that this is not necessarily true. The structural setting is the leading factor in the hydrogeology of this aquifer.

Another major difficulty inherent in karst is the high anisotropy. Keeping in mind the high cost of deep wells' drilling, the risk of failing to penetrate the water-bearing openings should be minimized. A schedule of cooperation between hydrogeologists and petroleum engineers is suggested to minimize the risks. Oil drillers should remove some restrictions concerning the data they obtain from the drillings and should include some sampling and measurements related to hydrogeology in their drilling schedule. Sampling the water formation not only at deep horizons but also at shallow depths is essential for hydrogeologists. The distribution of porosity, permeability, salinity, temperature, viscosity, and pressure with time and space indicates the hydrodynamic pattern of the aquifer. However, the hydrogeologist should justify his requests and assure the productive usage of the parameters he needs.

References

DSİ. (1970). Hydrogeological investigation report of Mardin-Kiziltepe plain, DSİ Press. Report No. II. 4/49-70, DSİ, Ankara, Turkey.

DSİ. (1987). Mardin-Nusaybin Beyazsu ve Karasu kaynaklarının geliştirme imkanlarının araştırılması karst hidrojeolojisi ön inceleme raporu, Ankara, Turkey (in Turkish).

Eroskay, S. O., & Günay, G. (1980). Tecto-genetic classification and hydrogeological properties of the karst regions in Turkey. in G. Günay (Ed.), Karst Hydrogeology Proceedings: October 1979, Oymapınar-Antalya, Turkey, UNDP Project TUR/77/015, p. 1–41.

Yalçın, N. (1976). Geology of the Narince-Gerger area (Adıyaman province) and its petroleum possibilities. Revue de la Faculte des Sciences de l'Universite d'Istanbul, serie B, 41, 57–82.

Yılmaz, Y. (1993). New evidences and model on the evoluton of the Southeast Anatolia orogen. Gegical Society of America, Bulletin, 105, 251–271.

Karst Hydrogeology of Muğla—Gökova Karst Springs

12

Türker Kurttaş, Gültekin Günay, and Ali Gemalmaz

Abstract

There are three main karst spring groups in the area: Gökova, Azmak, and Akbük Bay. Gökova springs discharge from the coastline in the north of Gökova Bay. Azmak spring group discharges through the fault line in the east of Akyaka village. Akbük Bay spring discharges from alluvium at an elevation close to the sea level, situated in the northwest of Akbük Bay.

12.1 Introduction

The Gökova Bay is located at the edge of the Western Taurus Mountains of SW Turkey. The Jurassic-Cretaceous age of limestone and Eocene-Miocene age of conglomerates are the karstified rocks in the area. The altitude rises from the sea level to 1200 m within 2 km to the land. There are also widespread Quaternary alluvium and springs discharging to the Bay at the front of the steep topography (Gemalmaz 1994; Kurttaş et al. 1997, 2000).

12.2 The Springs of Gökova

The springs of Gökova consist of the springs that are discharged from the coastline between Akyaka-Akbük in the north of Gökova Bay. The springs are related to the Gökova fault system. In the 60-km long, fault system located in the north of Gökova Bay, which is a graben, still in the process of opening, the southern blocks have descended while the northern blocks have arisen. Due to the contact with the faults from the Quaternary age, the limestones in Haticeana and Babadağ, and the underground system of Köprüçay conglomerates have descended (Fig. 12.1). Thus, the karst system and the alluvial sediments under the sea level were affected by the saltwater. The springs are recharged as they flow and pass through the saltwater zone along the fault lines from the north and northeastern. The spring that surfaces from underneath the coastal alluvium in the north of Akbük bay is the discharge from Haticeana limestones situated there, again in relation to the faults. It is presumed that in the west of Akbük bay, Yatağan formation situated on the block that has descended due to fault formation, constitutes a barrier before karstic limestones, and it has an effect on the formation of the springs (Eroskay 1992). Hydrogeologic map of Gökova Bay (Fig. 12.1).

12.2.1 Azmak Spring Group

Azmak spring group is constituted by the discharges that occur from the slope of the fault line, located along the road between Akyaka-Çaydere, in the east of Akyaka. The springs have surfaced at elevations near sea level. The channel into which they are discharged forms (Azmak) a group of springs. The measurements, made by State Hydraulic Works (DSİ) in August 1990, show the total flow rate of the spring group as 11.18 m^3/sec. It has been observed that the results of the salinity measurements applied to the spring group varied depending on the place and the time (Eroskay 1992) (Fig. 12.2).

12.2.2 The Akbük Bay Springs

The spring is discharged from alluvium at an elevation close to the sea level in the northwest of Akbük Bay. It has formed

T. Kurttaş
Department of Geology, Hacettepe University, Ankara, Turkey
e-mail: kurttas@hacettepe.edu.tr

G. Günay (✉)
Hacettepe University, Yaşamkent Mah. Ataşehir Sitesi 7A- 5. Blok No:12, Çankaya, Ankara, Turkey
e-mail: ggunay@ttmail.com

A. Gemalmaz
Turkcell Company, Maltepe, Turkey

G. Günay et al., *Caves and Karst of Turkey - Volume 2*, Cave and Karst Systems of the World,
https://doi.org/10.1007/978-3-030-95361-4_12

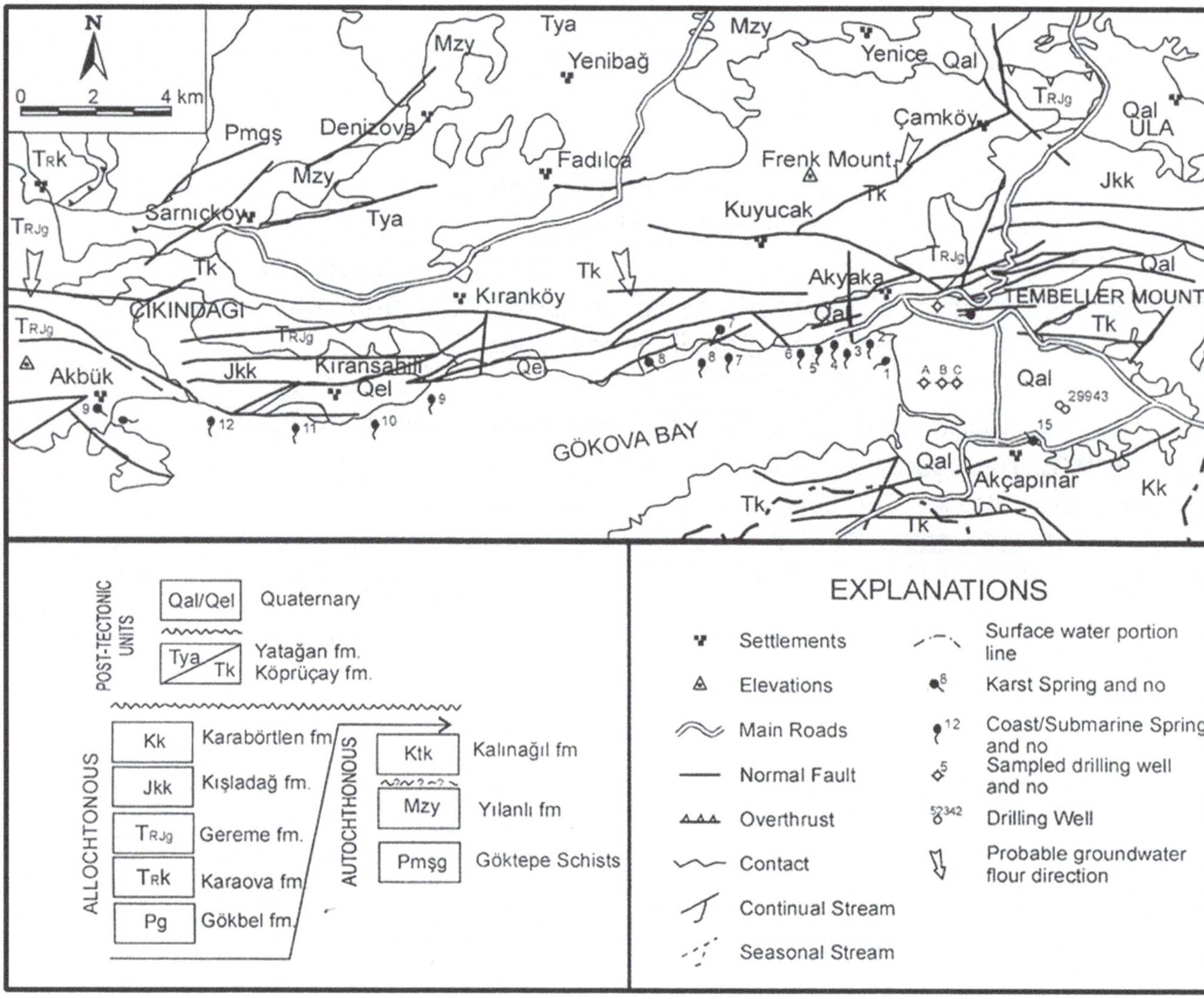

Fig. 12.1 Hydrogeological map of study area

300 m long Azmak. The spring is controlled by a fault. The Haticeana formation is situated behind the fault. Most probably, the spring is recharged by the Yılanlı and Haticeana limestones. According to measurements in July 1991, 20 °C of temperature, 0.007 of salinity, and 11,000 mho/cm of electrical conductivity (EC) were observed (Eroskay 1992) (Figs. 12.2 and 12.3).

12.3 Hydrogeological Characteristics of the Rocks

In the study area, the autochthonous units that form the Menderes Metamorphides, the allochthonous units that form the Elmalı Nappes, and the neo-autochthonous that cover them are stripped (Fig. 12.1). In determining the hydrogeological characteristics of the rocks, a previous study by Eroskay et al (1992), the water chemistry analysis results, the evaluations of the isotope data, and the field observations were used. Taking into consideration the geological positions and the lithological situations of the rocks, their practical permeability was discussed.

12.3.1 The Hydrogeological Characteristics of Autochthonous Units

The autochthonous units that are stripped in the study area, from the bottom to the top, are the Göktepe schist, the Yılanlı, and Kalınağıl formations. The Göktepe schists, mainly composed of clay schist, mica schist, chlorite schist, phyllite, quartzite, calc-schist, and marble lenses, are practically impermeable and serve as a hydrogeological barrier almost everywhere.

The Yılanlı formation located on the Göktepe schists is the most important unit in the area that shows aquifer

Fig. 12.2 Gökova springs in the Azmak spring Grouu area

characteristics. It is composed of limestone, dolomite, and marble. In the Muğla basin, the formation is intersected with 30–40-km long and 500–600-m thick young faults (Eroskay 1992). Due to these faults, several sinkholes and polje formations emerged. Many springs in the research area emerge and/or recharge from this unit. This unit is mainly collapsed and fractured due to the faulting and as a consequence, it has secondary porosity. It is hydrogeologically accepted as permeable.

The Kalınağıl formation which forms the upper part of the autochthonous sequence shows permeable characteristics with regard to the breccia limestone, red cherty, micritic limestones in the bottom, whereas it shows permeable characteristics with regard to the shale, marl, siltstone, and olistostromal formations that exist in the upper parts.

12.3.2 The Hydrogeological Characteristics of the Allochthonous Units

The formations that compose the allochthonous units, from the bottom to the top, are the Gökbel, Karaova, Gereme, Kışladağ, and Karaböğürtlen formations. The Gökbel formation, composed of dolomites and dolomitic limestones, forms the lowest level of the allochthonous sequence and rises in few areas. Since there is no important recharge area due to the narrow dissipating area, it is a hydrogeologically permeable unit.

The Karaova formation located on the Gökbel formation is impermeable due to its lithological structure composed of day Limestone at the bottom, and sandstone, pebble, siltstone, shale, marn, and mudstone at the upper parts.

The most important aquifer in this allochthonous sequence is the Gereme formation. The unit is composed of intensively karstified, highly collapsed, and fractured dolomitic limestones, recrystallized limestones, and marbles. The biggest discharges originate from this formation. It is in contact with the Yılanlı formation of the Menderes Massif. It is hydrogeologically permeable.

The Kışladağ formation, the bottom of which is composed of radiolarite and chert banded micritic limestones is practically impermeable. Yet, it is permeable where cemented calcareous breccia and micritic Limestone sequences are observed.

The Karaböğürtlen formation, which forms the uppermost layer of the allochthonous sequence, serves as a

Fig. 12.3 Muğla Akbük village and Akbük springs

hydrogeological barrier. The mentioned unit is composed of sandstone, siltstone, shale, and marn succession, and it is practically impermeable.

The Kertmeç mélange and Fethiye peridotites, which cover the allochthonous and autochthonous sequences, are also hydrogeological barriers. The Fethiye peridotites, composed of dunites, harzburgites, and pyroxenites, and Kertmeç melange that developed due to the settling of peridotites, include parties from the lower layers, sandstone, siltstone, day stone, and olistostromal levels.

References

Eroskay, S. O., Gözübol, A. M., Gürpınar, O. & Şenyuva, T. (1992). Muğla-Gökova ile Milas-Savran ve Ekinambarı karst kaynaklarının jeolojik ve hidrojeolojik incelemesi, Sonuç Raporu. DSİ Genel Müdürlüğü, Ankara (in Turkish).

Gemalmaz, A. (1994). Gökova karst kaynakları sisteminin uzaktan algılama ve coğrafi bilgi sistemi teknikleri ile değerlendirilmesi. Yüksek lisans tezi. Hacettepe Üniversitesi Fen Bilimleri Enstitüsü, Beytepe, Ankara, 95s., Ankara (in Turkish).

Kurttaş, T, Günay, G. & Gemalmaz, A. (1997). Denizaltı kaynaklarının uzaktan algılama teknikleri ile belirlenmesi: Gökova Körfezi örneği. Türkiye Ulusal Fotogrametri ve Uzaktan Algilama Birligi, 3. Uzaktan Algılama ve Türkiye'deki Uygulamaları Semineri. 16–18 Mayıs 1997, Uludag-Bursa (in Turkish).

Kurttaş, T., Günay, G. & Gemalmaz, A. (2000). Karst hydrogeology of the Gökova. In G. Günay (Eds.) Guide Book, Present state and future trends of karst studies, September 17–26, 2000, Marmaris-Turkey (ISBN:975-491-098-7).

Karst Springs and Waterfalls—Zamanti River, Eastern Turkey

13

Gültekin Günay and Göksel Övül

Abstract

The Zamantı River prevails as an erosion base level for all water discharges within the Kayseri Province area. It is one of the two branches that create the most important river in the region—the Seyhan River. The other branch is the Göksu River. The spring zone of the Zamantı River is located at the karstified formation of the Jurassic-Cretaceous and also Permian limestones. Only Serefiye karst springs emerge from Pliocene-aged karst limestones. Kaynar and Elbaşı karst springs emerge from Permian-aged karst limestones. The most prominent karstic springs that create this river are the Pınarbası, Serefiye, Tacin, Kaynar, and Elbası springs. Cumulative spring discharges vary (Tacin Stream) between Q_{min} 2 m^3/s, Q_{max} 34 m^3/s. Discharge regimes of all the springs are carefully monitored and analyzed including recession curve analysis and calculation of active storage volume of the broad karst aquifer. The catchment area of the Zamantı River, before the confluence with Göksu River, is 8700 km^2 with an average annual flow of 65,603 m^3. The rate of the river flow ranges between Q_{max} 970 m^3/s and Q_{min} 29 m^3/s. A few fascinating waterfalls are formed along the canyon (Zebil and Kapuzbasi waterfalls). The water, which creates the Kapuzbası waterfalls, discharges from seven close-spaced karstic springs emerging from cavern outlets. All springs and surface flows are under strict environmental protection as national heritage.

G. Günay (✉)
Hacettepe University, Yaşamkent Mah. Ataşehir Sitesi 7A- 5. Blok No:12, Çankaya, Ankara, Turkey
e-mail: ggunay@ttmail.com

G. Övül
Hacettepe University, Ankara, Turkey

13.1 Introduction

Kayseri–Pınarbaşı–Uzunyayla Basin is located 50 km east of Kayseri, on the Kayseri–Malatya State Highway. Its area is approximately 4375 km^2. In the north, Mt. Hınzır and Mt. Korumaz are located, while in south Aygörünmez, Soğanlı, and Gövdeli mountains are situated. The flat areas in the basin, which can be considered as plains, make up 1150 km^2, plateaus cover 850 km^2, and mountainous terrain is 2375 km^2. The average annual precipitation is 433 mm, and the average annual temperature is 7.7 °C. Pınarbaşı district and Pazarören and Kaynar subdistricts are the main settlement areas in the basin.

Three plains in the basin were found to be good for groundwater exploitation. These are Elbaşı–Çörümşek, Şerefiye–Viranşehir, and Kaynar–Tersakan plains.

The rocks which create aquifers in the basin are the Paleozoic-aged limestones and marbles, the Pliocene-aged lagoon limestone, and fragmental lagoon deposits.

The recharge of groundwater in the basin is through the precipitation and percolation from surface flows. Discharge occurs through springs and evapotranspiration.

Annual safe yield in the basin is 12.4 × 106 m^3/year at Elbaşı–Çörümşek Basin, 74.2 × 106 m^3/year at Viranşehir–Şerefiye Basin, 11.2 × 106 m^3/year at Kazancık Basin, and 9.3 × 106 × 106 m^3/year at Tersakan–Kaynar Basin.

The rock units in the basin were studied under three units due to facies and tectonic variations. Their correlation was partly done during the studies carried out in the western and central Taurus Mountains by geologists from MTA and the French geology team.

In the north of the basin is the Bolkardağ Unit; in the center, Bozkır Unit; in the south, rocks from Geyikdağ Unit are located. The rock units at Bolkardağ Unit are of the Permian and pre-Permian ages, and they are represented by green and blueschists, marble, and limestone.

The rocks of Bozkır Unit are Cretaceous-aged, and they are represented by ophiolites, submarine volcanic rocks,

G. Günay et al., *Caves and Karst of Turkey - Volume 2*, Cave and Karst Systems of the World,
https://doi.org/10.1007/978-3-030-95361-4_13

pelagic limestone, and radiolaria, as well as by allochthon rock units belonging to Bolkar and Geyikdağ units.

The rocks of the Geyikdağ Unit were formed in the Upper Devonian–Eocene period. They are represented by carbonate and fragmental rocks Ketin (1959a, b, 1976).

Continental climate is observed in the study area. Winter is long and with snowfall, spring is short and wet, and the summer is dry and hot. Data obtained from the Pınarbaşı Meteorological Station, which is located in the study area, were used. According to the data, the highest precipitation occurs in May with 57.6 mm and the lowest in July with 9.6 mm.

13.2 Geology

The rock units in the basin were studied under three units due to facies and tectonic variations, and the stratigraphy and tectonics of these units were dealt with separately. The correlation of the studies in the basin was partly achieved by MTA geologists who conducted comprehensive investigations in the western and central Taurus Mountains.

In the north of the basin is the Bolkardağ Unit; in the center, Bozkır Unit; and in the south, Geyikdağ Unit is located. The direction of the units is northeast and southwest.

Bolkardağ Unit consists of Permian and pre-Permian-aged limestone, marble, green, and blueschists. Green stones are located at the bottom and are covered by marble and limestone units, respectively. The base of the limestone is discordant. The base of the Bolkardağ Unit is observed in the west of Mt. Korumaz (outside the basin) between Büyük Bürüngüz village and Bünyan district. It was moved over the Cretaceous-aged flysch by a thrust fault. The Eocene is discordant over these.

Bozkır Unit includes the Cretaceous-aged rocks. It is composed of serpentine, diabase, spilite, durit, peridotite, and limestone olistolith of various ages, as well as red and green pelagic limestone, and radiolaria in the upper layers. Submarine volcanism and gravity tectonics were influential. It is covered by Eocene-aged deposits, which are discordant.

Geyikdağı Unit includes the rock units formed in the Upper Devonian–Eocene period. There is incongruity at Carboniferous, Permian, Triassic, and Jurassic bases. It is represented by carbonate and fragmental rocks. Metamorphism is not observed. The unit is covered by Eocene-aged deposits, which are discordant.

Pliocene is represented by white, khaki, and ash-colored pebbles, sandstone, clay, and limestone. It covers the Bolkardağ, Bozkır, and Geyikdağı units with Miocene-aged deposits.

13.3 Hydrology

The most important stream which receives the karst springs from within the borders of Kayseri Province is the Zamantı River. Being one of the two main branches of Seyhan River (Göksu and Zamantı Rivers), Zamantı River's springs are located in Şerefiye Village near the subdistrict of Örenşehir in the middle of Uzunyayla. Its length within the borders of Kayseri is 230 km. Springs outflow initially from Uzunyayla, Zamantı River. The first follows the north–south flow direction, and in the northeast of Pınarbaşı district, the flow direction shifts to the northeast–southwest. Along its course, Zamantı River flows through narrow and deep valleys, and after joining with the peripheral creeks, it flows out of the borders of Kayseri from the South of Yahyalı District. At the slopes of the Akinek Mountain near Aladağ District (80 km from Adnan), it joins Göksu River to make up the Seyhan River. With a precipitation area of 8700 km^2, Zamantı River's average annual flow rate is 65,603 m^3/year. The highest flow rate of the river is 970 m^3/s, and the water depth at this flow rate can go up to 410 ^{cm}. The minimum flow rate of the Zamantı River is 29 m^3/s. The Zamantı River is suitable for rafting, and the 10-km-long Zamantı Canyon, which also includes Zebil Waterfalls, is. Kapuzbaşı Waterfall located in the Yahyalı district is formed by seven karstic springs emerging from the karstic dissolution cavities in an area of 500 m. These springs are located within the borders of Kapuzbaşı village, 76 km out of the Yahyalı district. Besides Kapuzbaşı Waterfall, Elif, Yeşilköy, and Derebağ are among the waterfalls with high flow rates in the vicinity of the Yahyalı district.

Several karstic springs with high flow rates exist in the Uzunyayla region, from where the Zamantı River outflows. Among these are Pınarbaşı, Şerefiye, Tacin, Kaynar, and Elbaşı karstic springs. The hydrological and groundwater studies were conducted at these springs, and their recession curve coefficients were calculated. Active storage volumes of the karstic aquifers were also calculated (Evsen 2006).

The major and continuous surface waters of the basin are the Zamantı River, and Tacin, Çörümşek, and Elbaşı Creeks.

13.4 Hydrogeology

The aquifers in the basin are composed of Paleozoic-aged limestone–marble units, Pliocene-aged lagoon limestone and sandstone, pebbles, clay, and tufa units, respectively.

In the Pınarbaşı Uzunyayla Basin, the crystallized limestone–marble and dark-colored limestone units, which make up Mt. Hınzır and Mt. Korumaz and which belong to Bolkardağ Unit, have aquifer features. These units crop out in the north of the basin. Its total thickness reaches a few kilometers. Underneath the marble and limestone, there are metamorphic schists. With a primitive relationship, the bases of the marble and limestone, together with the schists, are not observed in the field. However, in the west of Mt. Korumaz, near Büyük Bürüngüz village, it is abnormally located (by thrust fault) on a different facies made up of presumably Upper-Cretaceous-aged Çarmaklı limestone, being afonitik limestone, sandstone, and serpentine. High productivity was achieved from the limestone unit at the exploitation wells located near Hazerşah-Süksün village in the outer north of the basin.

It is understood from the investigation and exploitation wells, which reach the marble and dark limestone representing the aquifers in the basin, that the hydraulic properties of the units are very good. The conductivity of the Permian and pre-Permian-aged limestone and marble at Akçalı well (No. 22112), Akören well (No. 12804), and at the exploitation wells drilled in the cooperative areas of Hazerşan-Süksün villages in the outer north of the basin, it varies between 115 and 2646, and their specific yield is between 0.6 and 36. The depths of the wells reaching this aquifer rock are around 300 m. The thickness of the unit is not certain. It was understood that its thickness may be up to a few kilometers (Göğer 1978). Study area map (Fig. 13.1), Hydrology map (Fig. 13.2), Kapuzbaşı karst springs (Figs. 13.3, 13.4, 13.5, 13.6, 13.7, 13.8, 13.9, 13.10 and 13.11).

13.5 Recession Curve Analysis of the Large Karstic Springs

Pınarbaşı-3 Springs: At the beginning of the spring recession period in 2004,

$Q_o = 963$ *l/s* = 83,203 m^3/day and

The flow rate on 01.03.2005 which is 240 days after 30.06.2004 is $Q_t = 680$ *l/s* = 58,752 m^3/day.

When in the $Q_t = Q_o \times e^{-\alpha t}$ equation, these values and $t = 240$ are put into place, and algorithms of both sides are taken;

Fig. 13.1 Study area of zamanti river

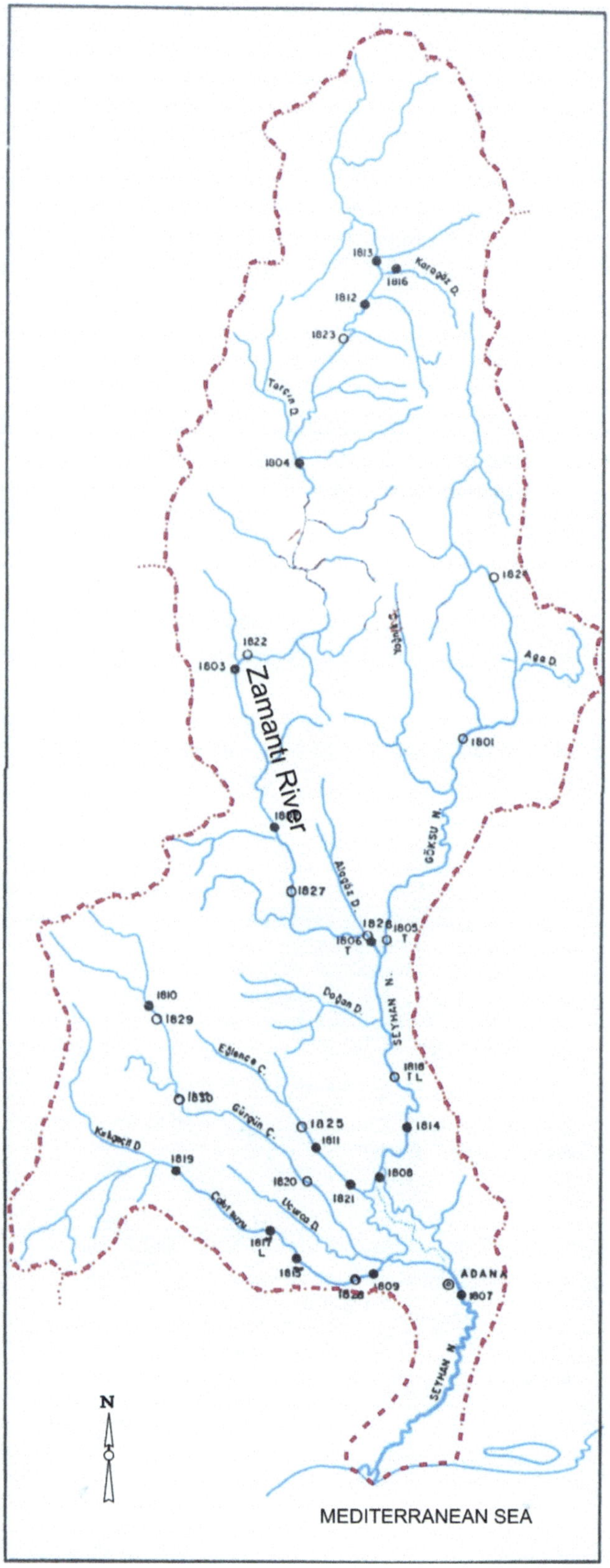

Fig. 13.2 Hydrologic map of the zamanti river area

$$\log 58,752 = \log 83,203 - \propto \times 240 \times \log 2.71828$$

$$\propto = (4.92 - 4.77)/104.232$$

$$\text{Recession Coefficient } \propto = 0.00143\text{day}^{-1}$$

$V_s = \int Q_t \times dt = Q_o / \propto$ equation gives the amount of water flowing out from the spring in time t, and at the same time, the volume of the water stored by the spring.

When Qo ve ($\propto$) values are put into place;

Vs = 83,203 / 0.00143 = 58.2 × 10^6 m^3/year is found.

($\propto$) recession coefficient is a parameter which depends on the geology and morphology of the spring zone. While high values indicate the greatness of transmissivity value, they also show the smallness of the discharge area. Moreover, the ($\propto$) recession coefficient value found above ($\propto$ = 0.00143 day^{-1}) shows that the groundwater is developing along fractures.

Şerefiye-2 Springs: At the beginning of the spring recession period (29.04.1969) in 1969, the flow rate is Q_o = 10,836 l/s = 93,623 m^3/day, and on 30.12.1969 which is 240 days after 29.04.1969, it is Q_t = 2067 *l/s* = 178,588 m^3 day.

When in the $Q_t = Q_o \cdot e^{-\propto t}$ equation, these values and t = 240 are put into place, and algorithms of both sides are taken;

$$\log 178,588 = \log 936,230 - \propto \times 240 \times \log 2.71828$$

$$\propto = (5.97 - 5.25)/104.232$$

$$\propto = 0.007 \text{ day}^{-1}$$

$V_s = \int Q_t \times dt = Q_o / \propto$ equation gives the amount of water flowing out from the spring in time t, and at the same time, the volume of the water stored by the spring.

When Q_o ve ($\propto$) values are putinto place;

V_s = 936,230/0.007 = 133,7 × 10^6 m^3/year is found as the volume of water stored by the spring. While high values of ($\propto$) coefficient indicate the greatness of transmissivity, they also show the smallness of the discharge area. The $\propto$ = 0.007 coefficient found above shows that the groundwater is developing along fractures.

Yukarı Karagöz-2 Springs: At the beginning of the spring recession period (08.03.1998) in 1998, the flow rate is Q_o = 2455 *l/s* = 21,212 m^3/day.

And on 08.03.1999 which is 240 days after 08.03.1998, it is Q_t = 1716 *l/s* = 148,262 m^3/day.

When in the $Q_t = Q_o \cdot e^{-\propto t}$ equation, these values and t = 240 are put into place, and algorithms of both sides are taken;

Fig. 13.3 Kayseri-Yahyali Kapuzbaşi Karst spring for waterfall

$$\log 148,262 = \log 212,112 - \propto \times 240 \times \log 2.71828$$

$$\propto = (5.326 - 5.17)/104.232$$

$$\propto = 0.00149\ \text{day}^{-1}$$

$V_s = \int Q_t \times dt = Q_o / \propto$ equation gives the amount of water flowing out from the spring in time t, and at the same time, the volume of the water stored by the spring.

When Q_o ve ($\propto$) values are put into place in V_s equation, V_s = 212,112/0.00149 = 142 × 10^6 m^3/year is found.

Fig. 13.4 Kayseri, Kapuzbaşi Karst springs

Fig. 13.5 Kayseri Kapuzbasi Karst springs

Fig. 13.6 Kayseri Kapuzbaşi karst springs

Fig. 13.7 Kayseri Kapuzbaşi Karst springs

Fig. 13.8 Big Karst springs Kapuzbaşi

Fig. 13.9 Karst springs waters at Kapuzbaşi area

Fig. 13.10 Kayseri-Yahyali Kapuzbaşi Karst springs

Fig. 13.11 Dr. Günay in working area

13.6 Conclusion

Recession coefficient is a parameter which depends on the geology and morphology of the spring zone. While high values of (∝) coefficient indicate the greatness of transmissivity, they also show the smallness of the discharge area (Karanjac and Günay, 1980; Günay and Yayan, 1979). The ($\propto = 0.00149$ day$^{-1)}$ value indicates that development of groundwater flow is very well.

References

Evsen, A. (2006). Kayseri-Sarız Havzası Hidrojeolojik Etütü Planlama Raporu, DSİ 12. Bölge Müdürlüğü, Kayseri.

Göğer, E. (1978). Kayseri-Pınarbaşı ve Uzunyayla Havzası Planlama Kademesi Hidrojeoloji Raporu, DSİ 12. Bölge Müdürlüğü, Kayseri.

Karanjac, J., & Günay, G. (1980). Dumanlı Spring Turkey - the largest karstic spring in the world, Journal of Hydrology, 45, p. 219–231.

Günay, G., & Yayan, T. (1979). Antalya - Kırkgöz kaynakları hidrojeoloji incelemesi, 1. Ulusal Hidrojeoloji Semineri, DSİ Oymapınar Barajı, Antalya, DSİ-UNDP Projesi TUR / 77/ 015 Project Studies. DSİ Groundwater Dept. Yücetepe, Ankara.

Ketin, I. (1959a). The Orogenic Evolution of Turkey. MTA Publication, Ankara, Turkey.

Ketin, I. (1959b). Türkiyenin Orojenik Gelişmesi, MTA Yayını, Ankara, Türkiye.

Ketin, I. (1976). Main Orogenic Eevents and Paleogeographic Evolution of Turkey, MTA Publication, Ankara, Turkey.

Niğde–Pozanti Şekerpinari Springs, South of Turkey

14

Gültekin Günay

Abstract

The Şekerpınarı springs are in the Central Taurus Mountain region. The Şekerpınar spring is located in the Ecemiş Fault Corridor that separates the central and eastern parts of the Taurus Belt. The Şekerpınarı spring's discharge is around 6–12 m^3/s.

14.1 Introduction

The Şekerpınarı spring is located in Alpa town, Pozantı county on the Ankara-Adana highway and railway.

The geological survey of the research area will be defined in detail. The lithological structures will be distinguished, and the underground extensions, borders, contact relations, the lithological, and structural relations that can affect the groundwater circulation will be determined.

14.2 Geological Structure

The Bolkar Mountains form the backbone of the eastern part of the Inner Taurus belt which is mainly composed of slightly metamorphic limestones and slates of the Bolkar Group. The age of the Bolkar Group ranges from the Permian to Late Cretaceous. These rocks were cut by dibasic intrusions.

The formations of the Bolkar Group are thrust over formations of the late Cretaceous Paleocene age of the Eregli-Ulukışla Basin from south to north. The up-thrust can be observed from Karıncadağ from the east.

A different tectonic assemblage, Aladağ unit, (Özgül 1976) was differentiated in the south of Bolkar Mountains.

Thick Jurassic-Cretaceous carbonates (Cehennemdere Formation) unconformably overlie the Karagedik or Öşün formation. It is overlain by Late Cretaceous flysch containing ophiolitic olistostoststroms and olistolites (Aslanköy formation) tectonically overlain by ophiolites.

14.3 Hydrology

The Bolkar Mountains in the Central Taurids form the ENE-WSW directed elevations between 1500 and 3524 m.

The climate in the region, due to the high topography (1000–3500 m), shows relatively more continental characteristics compared to the Mediterranean climate of the coastal region. The summers are hot and dry. The winters are cold and wet. Most of the precipitation falls in the form of snow and at high elevations, and the snow cover stays even until the early summer. On the northern slopes of the Bolkar Mountains, the snow does not melt even in summer. The average precipitation between the 1980–2000 years is 678 mm/year (Figs. 14.1 and 14.2).

The Şekerpınarı spring's discharge is around 6–12 m^3/s. The springs regression curve analysis was prepared, and a coefficient is $a = 0.0178$ day^{-1}. The storage capacity of the spring is $V = 38{,}831 \times 106$ m^3 (Figs. 14.3 and 14.4).

Diving Team Studies to the Source to Research in the Şekerpinari Karst Springs

Şekerpınarı spring diving studies to investigate the hydrological condition of the Şekerpınarı resource diving activities to the source were carried out in two stages. The first time for the research of the Şekerpınarı spring was made in December 2000. The second time to Şekerpınarı spring was held between October 30, 2003, and November 7, 2003. During this specified period, divers dived to a

G. Günay (✉)
Hacettepe University, Yaşamkent Mah. Ataşehir Sitesi 7A- 5.Blok No:12, Çankaya, Ankara, Turkey
e-mail: ggunay@ttmail.com

G. Günay et al., *Caves and Karst of Turkey - Volume 2*, Cave and Karst Systems of the World,
https://doi.org/10.1007/978-3-030-95361-4_14

Fig. 14.1 Şekerpinari springs in Pozanti Area

depth of 40 meters and examined the first and second springs. In the second source, fishes of which many species could not be detected and whose length reached 40 cm were identified.

The scientist and diving team consist of a team of Serbian people.

A detailed report on diving works is available in the authors. In the study of the diving team, the following issues were determined.

- To perform spelodiving investigations along the accessible section of Sekerpınarı-1 and Sekerpınarı-2 karst

Fig. 14.2 Şekerpinari springs in Pozanti Area 2

channels. Those investigations include disassembling of the filter at the distance of 25 m from channel inlet in the Şekerpınarı-1.

- Collecting the water samples from different sections of the karst channels for turbidity analysis.
- Sampling of deposits from the bottom of the karst channels to estimate the quantity of suspended material (possible cause of turbidity).
- Collecting geological data along with the investigated parts of the karst channels.

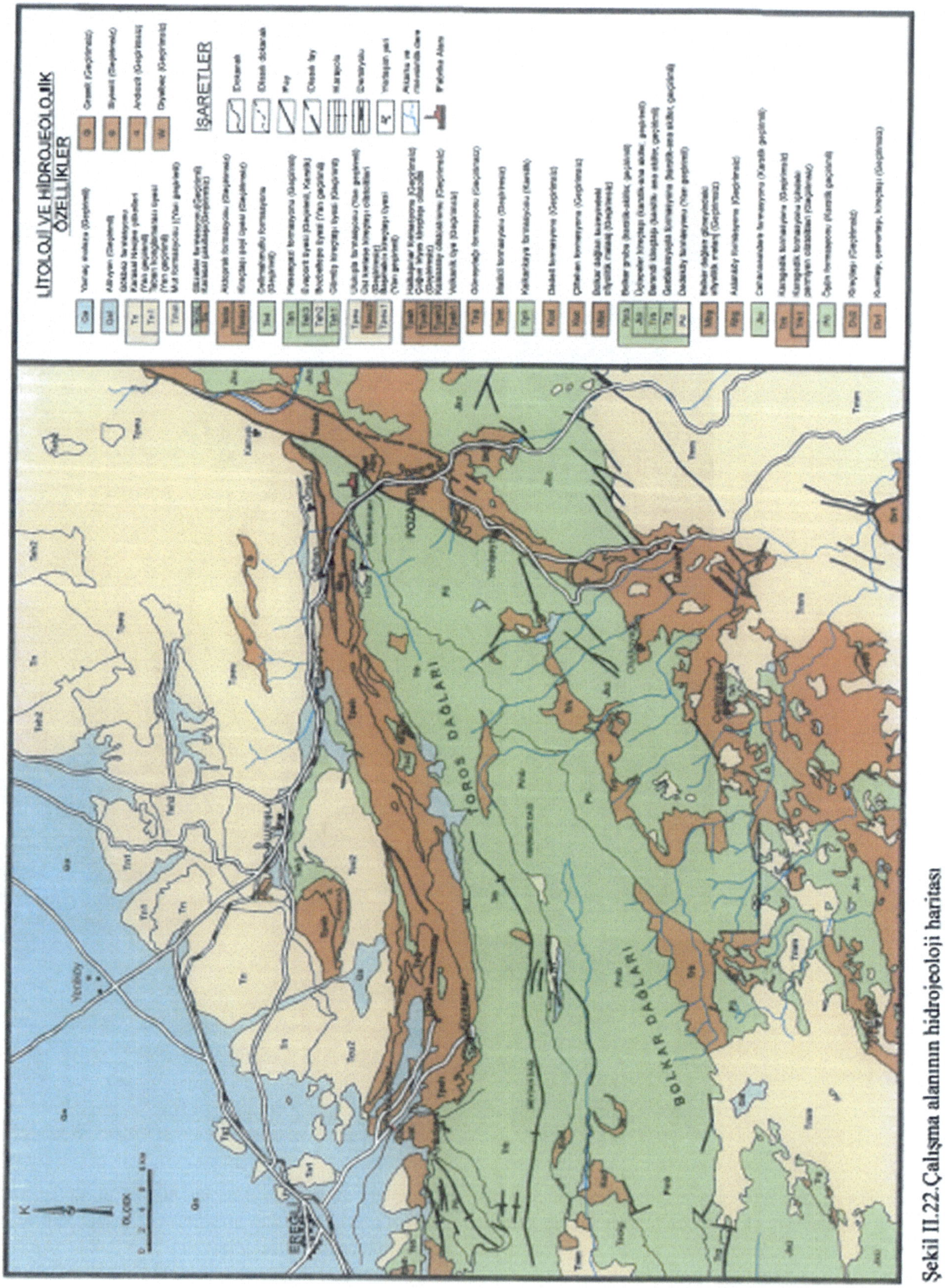

Fig. 14.3 Hydrogeologic map of the Şekerpinari spring area

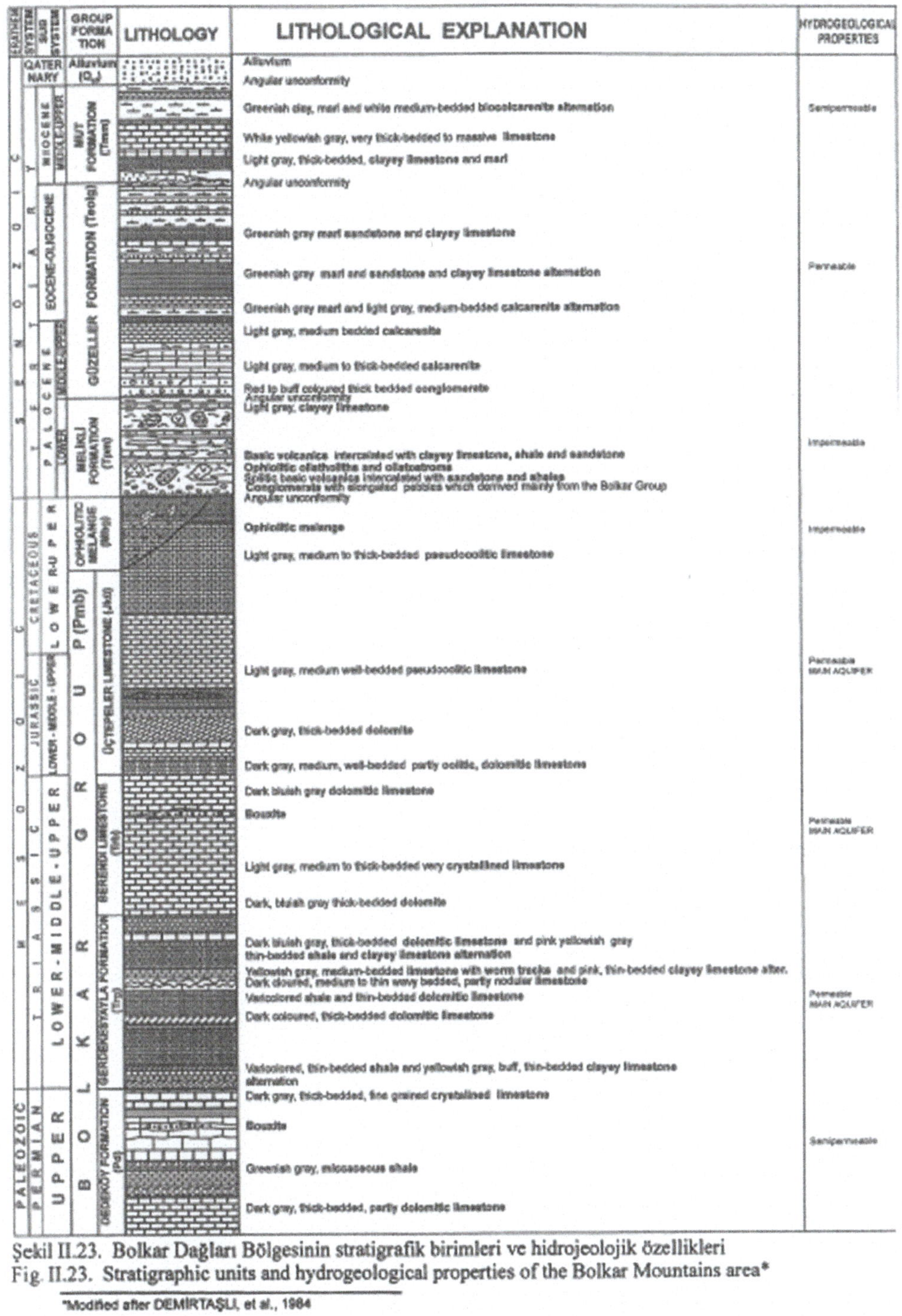

Fig. 14.4 Stratigraphic units and hydrogeological properties of the Bolkar Mountains Area

- Recording by underwater digital video camera all important stages of investigations and particularly possible locations for groundwater sampling.
- To propose location and kind of new tapping structure improvement of the existing structure.

For the successful execution of the above-listed topics to ultimate prerequisite was the minimal discharge of Sekerpınarı spring.

For a successful speleodiving, the minimum velocity of groundwater flow through the investigated karst channels should be less than 0.5 m^3/s (Figs. 14.5 and 14.6).

Diving Team

Prof. Dr. Gültekin Günay—Hydrogeologist—Hacettepe University.

Prof. Dr. Petar Milanovic—Hydrogeologist—Belgrad. Serbia.

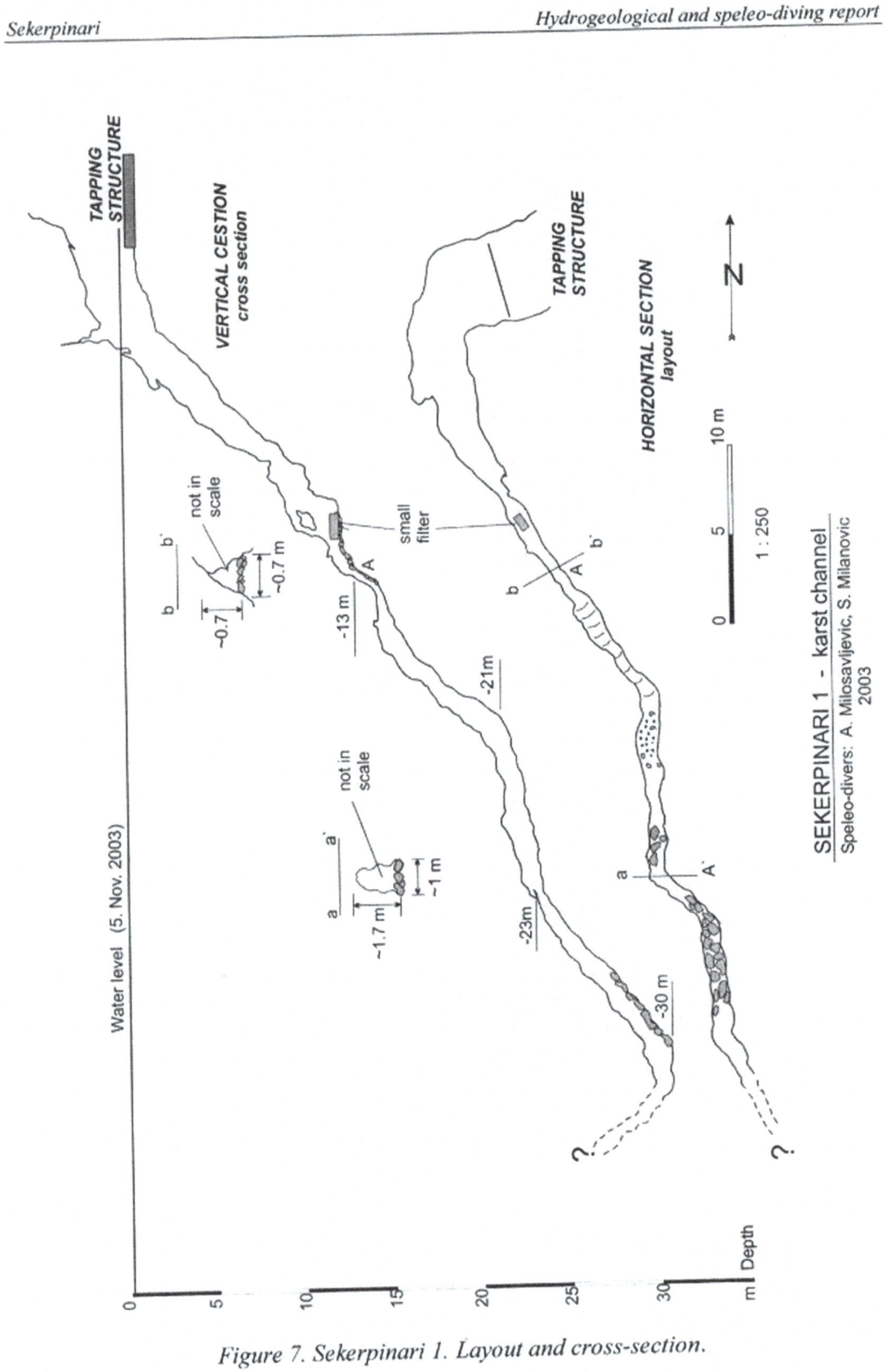

Fig. 14.5 Şekerpinari.1, karst channel

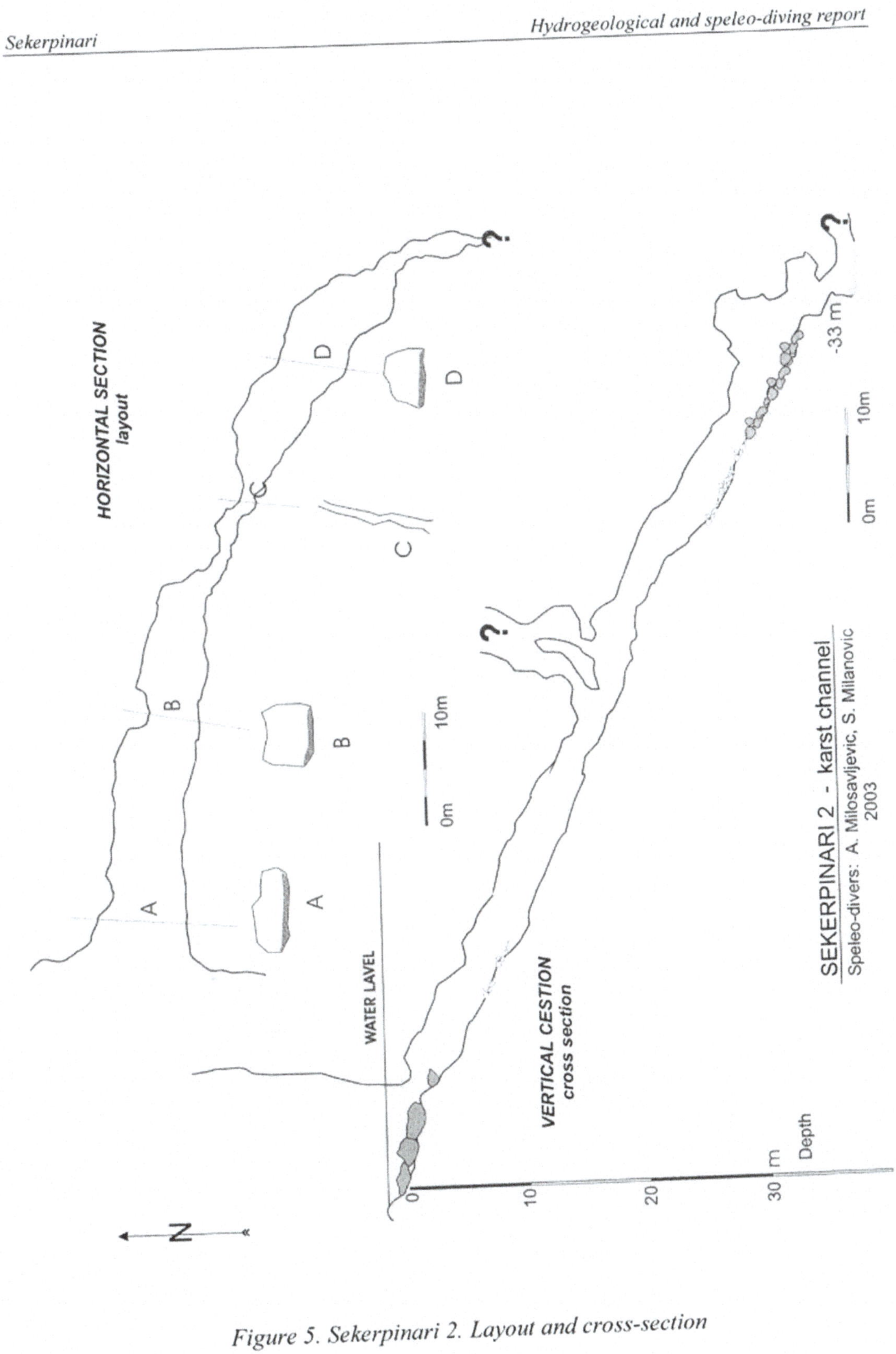

Fig. 14.6 Şekerpinari-2, karst channel

Dr. Sasa Milanovic—Hydrogeologist—Diver—Belgrad. Serbia.

Aleksandar Milosavijev—Professional Diver—Belgrad. Serbia.

14.4 Water Chemistry

In the Pozantı region, the values of calcium were measured. It is between 20.80 and 120.24 mg/I in spring waters and between 24.00 and 43.00 mg/I in Şekerpınarı. The values of magnesium in Şekerpınarı are 4.75–10.90 mg/I. Sodium 1.80 mg/I, Potassium 0,29 mg/I, Bicarbonate 109.80 mg/I, Sulfates 0.80 mg/I, Chloride 1.40 mg/I, pH: 7.9; TDS: 130 mg/I, EC: 200 mS/cm, Temperature: 9.0 °C (Günay 2000).

References

Akay, E., Uysal, Ş. (1988). Orta Toroslar'ın post Eosen tektoniği. MTA Dergisi, 108, 57–68 (in Turkish).

Demirtaşlı, E. (1981). Summary of the Paleozoic stratigraphy and Variscan events in the Taurous Belt. Newsletter, IGCP Project No. 5, Correlation of Variscan and Prevariscan Events in the Alpine Mediterranean Belt, 3 (1), 44–57.

Günay, G. (2000). The project of hydrogeology and environmental effects study of the recharge area of Pozantı – Şekerpınarı springs, Hacettepe University, Danone S.A. Danone Food and Drink Industry and Trading Joint Company – Hacettepe University UKAM. 2001, Final Report.

Özgül, N. (1976). Toroslar'ın bazı temel jeoloji özellikleri. Türkiye Jeol. Kur. Bült., 19 (1), 65–78 (in Turkish).

Karstic Hot Water Aquifers in Turkey

15

Şakir Şimşek

Abstract

The purpose of this paper is to determine the common properties of karstic hot water aquifers based on hydrogeology and drilling results of selected geothermal fields of Turkey. Over the last 50 years in Turkey, numerous geothermal fields were discovered, and different types of hot water aquifers (geothermal reservoirs) were identified in these fields. The hydrogeological properties of hot water aquifers such as lithology, thickness, and horizontal extension have an important role in geothermal energy production. Most of the hot water aquifers in Turkey are Mesozoic-aged karstic limestones and Paleozoic-aged marbles. Karstic hot water aquifers are found in the Sakarya-Kuzuluk, Yozgat-Bogazlıyan, Aydın-Salavatlı, Konya-Ilgın, Izmir-Çeşme, Ankara-Melikşah, Denizli-Pamukkale, and Afyon geothermal fields. These fields are characterized by high productivity, medium-to-low enthalpy, and geothermal fluids of a high CO_2 content that creates a scaling problem.

15.1 Introduction

As a result of exploration studies, many hot water aquifers have been discovered by the General Directorate of Mineral Research and Exploration of Turkey (MTA). The exploration studies include prospecting, hydrogeology, geochemistry, geophysical surveys (gravity, resistivity, and seismic), drilling, testing, and pre-feasibility studies. Many technical experiments have been carried out on geothermal energy based on the results of these activities.

Upon completion of these preliminary studies, drilling was done in approximately in 40 geothermal fields. The aquifers of 75% of these fields are composed of carbonate rocks and mainly limestone. The important calcareous hot water reservoirs and their respective discovery dates are Denizli-Kızıldere (1968), Afyon-Ömer, Gecek (1970), Izmir-Çeşme (1974), Ankara-Melikşah (1975), Sivas-Sıcakçermik (1976), Konya-Ilgın (1977), Kırşehir-Terme (1987), Ankara-Haymana (1987), Sakarya-Kuzuluk (1987), Çankırı-Çavundur (1988), Aydın-Salavatlı (1988), Yozgat-Bogazlıyan (1989), and Kütahya-Yoncalı (1989).

15.2 General Properties of the Karstic Hot Water Aquifers

Soluble rocks such as carbonates (limestone, dolomite, and dolomitic limestone) and sulfates (gypsum and anhydrite), saline rocks (NaCl and KCI) can be karstified although these features are more common in limestone. Widespread carbonate outcrops are found in the Alpine-Himalayan orogenic belt, and one-third of Turkey is covered by carbonate rocks (Eroskay and Günay 1980). Other countries such as China, Algeria, Czech, Slovakia, Switzerland, Russia, Belgium, former Yugoslavia, Bulgaria, Greece, Italy, Germany, France, Hungary, and Israel also have karstic hot water aquifers (Boldizsar 1976; Garagunis 1976; Shaterev 1976; Antonenko and Mavritsky 1978; Haenel 1985; Quilong et al. 1985; Arad 1988; Luo et al. 1988; Zhang 1988; Fekraoui 1990; Franko et al. 1990; Konokov and Drokov 1990; Rybach and Hauber 1990; Vandelberghe 1990).

The general properties of karstic hot water aquifers are similar to those of cold water aquifers. But hot water aquifers have a higher dissolution capacity due to their chemical composition, high temperature, and pressure. There is an important relationship between temperature and hydrothermal cave development (Ford 1988). As an interpretive result of the hydrogeological studies, it can be said that the main common properties of the karstic hot water aquifer in geothermal fields are as follows: high productivity (10–300 l/s for each well), characterizing the medium

Ş. Şimşek (✉)
Engineering Faculty-Beytepe, Hacettepe University, Ankara, Turkey
e-mail: ssimsek@hacettepe.edu.tr

G. Günay et al., *Caves and Karst of Turkey - Volume 2*, Cave and Karst Systems of the World,
https://doi.org/10.1007/978-3-030-95361-4_15

(0–150 °C) and low enthalpy fields (30–70 °C), geothermal fluids with a high CO_2 content that create a scaling problem and travertine deposits around hot spring sites.

The analyses of karstic hot springs and water from wells of geothermal fields show the high Ca++ and HCO_3 values as their main characteristics as follows:

$$Ca > Na \text{ or } Mg > K \text{ and } HCO_3 > SO_4 \text{ or } Cl$$

Because of the seawater intrusion at the Izmir-Çeşme field and the effect of evaporitic rocks at the Yozgat-Bogazlıyan field, these fields differ from the rest.

Lithology and tectonic movements serve as the main controls on karstification in crystalline limestones and marbles from the Paleozoic to the Cenozoic age in Turkey. There is a large stratigraphic gap between aquifers and impervious caprock units. Heating systems via heat flow have developed as a result of neotectonic activity, and geothermal fields have occurred along the active tectonic zones especially on graben structures. One example is the well-known touristic area of the Denizli-Pamukkale geothermal field. According to the hydrogeological model, the main hot water aquifers in this area consist of limestone and marble with high permeability characterized by a network of fractures, faults, and karstic features. The alternation of Pliocene claystone, marl and sandstone, and Paleozoic schists acts as an impermeable cap rock in the region. The recharge rate in the area is quite large due to the permeability of the rocks forming horsts and grabens. The recharge is mainly from the surface and underground waters originating at rainfall precipitation and infiltrating the basin.

15.3 Important Karstic Hot Water Aquifers and Their Classification According to Age

Important karstic hot water aquifers and their classification according to age are given in Table 15.1.

15.3.1 Cenozoic Limestone Formations

Karstic features were determined by a hydrogeological survey in the region (Canik 1981). The flow rate of the well is 110 l/s at 42 °C from compressor testing. In the Denizli-Kızıldere geothermal field, Pliocene-aged Sazak limestone forms a reservoir that is partly karstic. From the first well, steam and hot water mixture were obtained from a depth of 540 m at temperatures of 198 °C. Also, Paleozoic-aged marble formation in this region displays karstic properties in the same field. Karstic structures are determined through the bottom level of the Miocene limestone in the MH-1/A well drilled to a depth of 594 m in the Ankara-Melikşah geothermal field (Fig. 15.1) (Kartal and Keskin, 1976). According to the first results from the well, a flow rate of 300 l/s was obtained at 39 °C output temperature.

The karstic Eocene limestone formations, which are widespread in central and eastern Anatolia are reservoirs of hot water (Fig. 15.2). In the Yozgat-Bogazlıyan geothermal field, two wells were drilled (BB-I and BB-2) with flow rates of 125 and 100 1/s, respectively, at temperatures of 32–46 °C (Özmutaf and Yüce 1989).

15.3.2 Mesozoic Crystalline Limestone Formations

Some of the low-temperature geothermal aquifers consist of Mesozoic crystallized limestone formations. One of the hot water aquifers is the Lower Triassic limestone at a depth of 282 m in the Izmir-Çeşme geothermal field (Şahinci, 1988). In this well, 42 l/s of hot water was produced by pumping it at a temperature of 56 °C.

Another important Upper Cretaceous-aged hot water limestone aquifer was discovered in the Ankara-Haymana field. The H-4 well was drilled to a depth of 221 m and produced 52 l/s at a 45 °C temperature.

Table 15.1 Age of selected Karstic aquifer formations in Turkey

CENOZOIC	
Quaternary	Karahayıt[a], Bursa
Neogene	Denizli-Kızıldere[a] - one aquifer, Konya-Ilgın[a] -partly
Tertiary	Ankara-Melikşah[a]
Eocene	Yozgat-Boğazlıyan[a]
MESOZOIC	Izmir-Çeşme[a], Ankara-Haymana[a], Sakarya-Kuzuluk[a], Diyarbakır-Çermik[a], Çankırı-Çavundur[a], Amasya-Hamamözü, Eskişehir-Inönü
PALEOZOIC	Sivas Sıcak Çermik[a], Denizli-Kızıldere[a], Denizli-Pamukkale, Afyon-Ömer[a], Gecek[a], Heybeli[a], Manisa-Urganlı, Kütahya-Yoncalı[a], Kırşehir-Terme[a], Kütahya-Simav[a], Aydın-Salavatlı[a], Bursa

[a] Determined by drilling

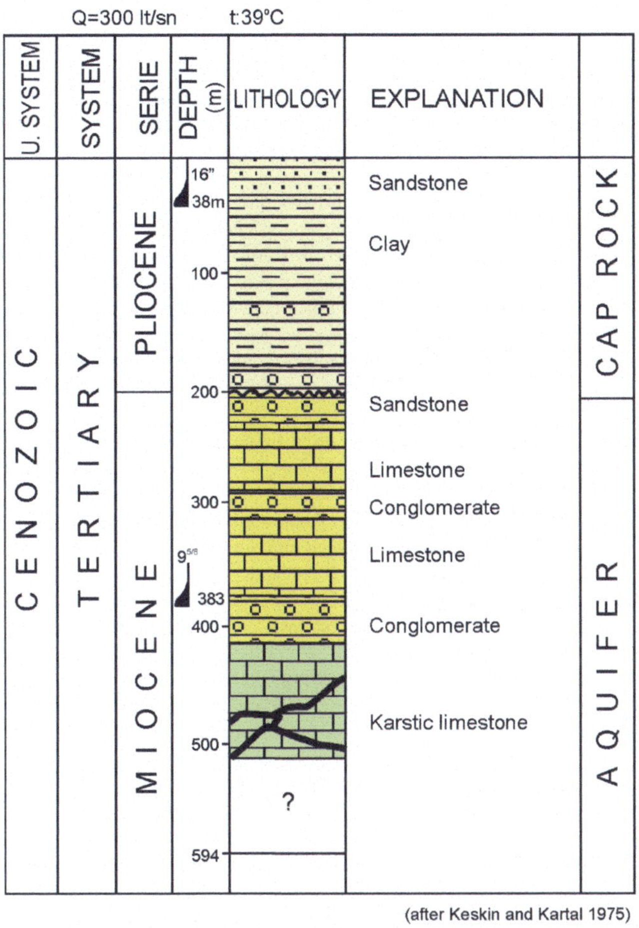

Fig. 15.1 Lithological log of Ankara-Melikşah Well No. MH-1A

15.3.3 Paleozoic Marble Formations

In many of the geothermal fields, karst is developed in the marble of the Paleozoic age. In both the Konya-Ilgın and Denizli-Kızıldere geothermal fields, Neogene limestone and Paleozoic marble form these geothermal reservoirs (Canik 1981; Şimşek 1985). In Aydın-Salavatlı geothermal field, a marble formation has flow rates of 300 t/h at each well (AS-I and AS-2). The wells are 1500 m and 962 m deep with temperatures at 162 °C and 171 °C, respectively. In the Kırşehir-Terme field, a marble formation is well-developed as a karstic aquifer, and hot water is produced from a depth of 333 m with a flow rate of 45 t/h (Özgür et al. 1987). In the Sivas-Sıcakçermik field, a total flow rate of 400 t/h hot water was produced from three shallow wells from depths of 200–350 m. Other important fields are Aydın-Germencik (232 °C), Kütahya-Simav (162 °C), Afyon-Ömer-Gecek (98 °C), and Heybeli (56 °C).

15.4 Exploitation Problems in Karstic Hot Water Aquifers

Among the factors affecting the production of hot water and steam, the most important problem is scaling. The most common is the deposition of $CaCO_3$ causing a reduction in the diameter of production and transmission pipelines and a drawdown in the reservoir. To solve this problem, downhole heat exchangers and chemical inhibitors are applied to the wells. At present, downhole heat exchangers are used in the

Fig. 15.2 Well logs are from Boğazliyan District in Yozgat City. **a** Lithological log of Yozgat-Boğazliyan Well No. BB. 1. **b** Lithological log of Yozgat-Boğazliyan Well No. 88.2 (After Özmutaf and Yüce 1989)

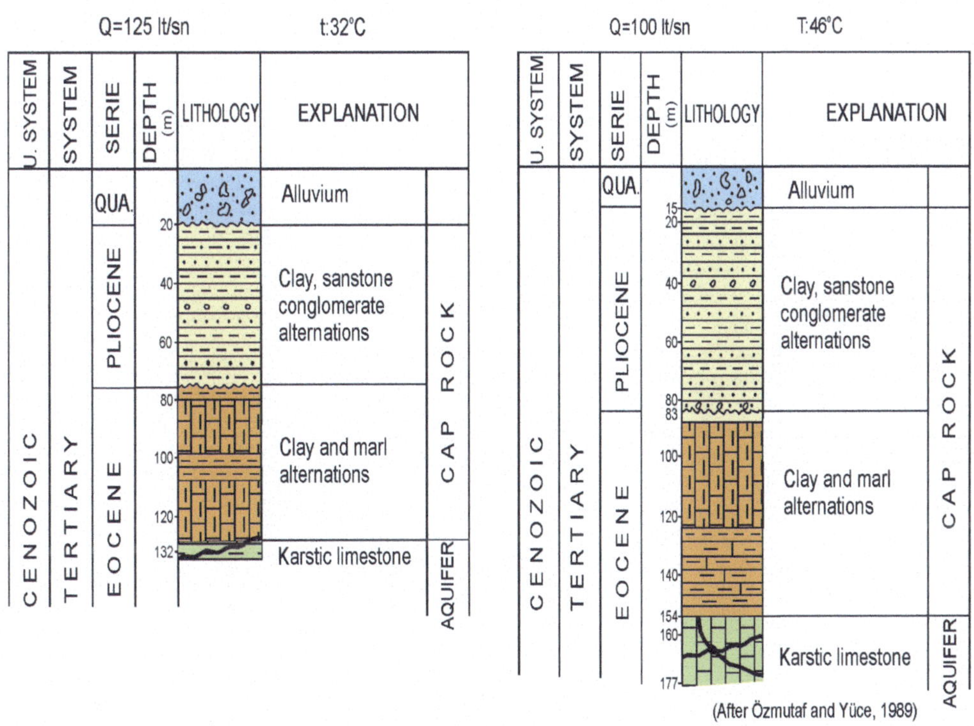

Afyon-Ömer, Sakarya-Kuzuluk, and Kütahya-Simav geothermal fields. In the Denizli-Kızıldere, Afyon-Ömer Gecek, Gazlıgol and Manisa-Salihli fields, chemical inhibitors have been successfully used. That way, this important problem encountered in karstic-type geothermal reservoirs during commercial usage can be eliminated.

Another major problem that might be faced during drilling is the mud leakage into the formation rendering the mud cement-control sometimes impossible. Therefore, if there are two karstic zones, only the shallower, low-temperature aquifer can be used, instead of the high temperature and high flow rate deeper aquifer. Cross section of well logs (Figs. 15.1 and 15.2).

15.5 Results and Suggestions

Important characteristics of karstic geothermal reservoirs in Turkey are as follows:

- they have a high production rate,
- they generally form low and medium enthalpy geothermal fields,
- in a high enthalpy field, they produce a high rate of CO_2 for commercial usage, as is the case at the Denizli-Kızıldere field,
- scaling in well-bore and transmission pipelines may pose a problem although this can be resolved by using downhole heat exchangers and chemical inhibitors.

Due to their high production rates, karstic reservoirs can be deployed for generating electricity, heating, and balneological and touristic purposes.

References

Antonenko, G. K. & Mavritsky, B. F. (1978). The geochemical features of thermal and mineral groundwater of the Rionion artesian basin and perspectives of their utilization. Proceeding of Hydrogeochemistry of Mineralized Waters Conference (Cieplice Spa, Poland), 263–270, Warsaw, Poland.

Arad, A. (1988). B, F and Sr as tracers in carbonate aquifers and in karstic geothermal systems in Israel. In: Karst Hydrogeology and Karst Environment Protection (Proc. 21st Congress of IAHS, Guilin, China, October 1988), vol. 2, 922–933. IAHS Publ. no. 176.

Boldizsar, R. (1976). Production of energy from the Pannonian geothermal anomaly. In: Proceedings of the Int. Congress on Thermal Waters, Geothermal Energy and Volcanism of the Mediterranean, vol. 1, 115–129. Athens.

Canik, B. (1981). Konya-Ilgın sıcak su kaynaklarının hidrojeoloji incelemesi. S.U. Fen Fakültesi Dergisi Sayı 1. seri A. s. 1–18, Konya (in Turkish).

Eroskay, S. O., & Günay, G. (1980). Tecto-genetic classification and hydrogeological properties of the karst regions in Turkey. in G. Günay (Ed.), Karst Hydrogeology Proceedings: October 1979, Oymapınar-Antalya, Turkey, UNDP Project TUR/77/015, p. 1–41.

Fekraoui, A. (1990). Outline of geothermal features in Algeria. Geothermal Resources Council Transactions, 14(l), 3–8, Hawaii, USA.

Ford, D. (1988). Characteristics of dissolution cave system in carbonate rocks. In: Paleokarst, 25–36. Springer-Verlag, New York.

Franko, O., Bodis, D., Fendek, M. & Remsik, A., (1990). Outline of geothermal activity in Czechoslovakia. Geothermal Resources Council Transactions, 14(l), 31–40, Hawaii, USA.

Garagunis, C. N. (1976). Geothermal areas in the Magic Gulf region in Greece. In: Proceedings of the Int. Congress on Thermal Waters. Geothermal Energy and Volcanism of the Mediterranean Area, vol. 1, 241–274, Athens.

Haenel, R., (1985). Present status on utilizing geothermal energy in the Federal Republic of Germany. In: Int. Symp. on Geothermal Energy (Hawaii, USA), 69–77.

Kartal, T. & Keskin, B. (1976). Ankara ili Çubuk kazası Melikşah jeotermal enerji alanina ilişkin arama sonucu, MTA Derleme No: 5737, Ankara (in Turkish).

Konokov, V. I. & Drokov, I. M. (1990). Utilization of geothermal energy USSR Geothermal Resources Council Transactions, vol. 14, Part l, Hawaii, USA.

Luo Xiangkang, Li Hongjue, Jiang Bing, Lu Duanfu, Wu Wenxun, Lin Shigui & Wang Yujie. (1988). The characteristics and resource assessment of the thermal groundwater from the Lower Triassic series karst in Chongqing Xiaoquan Hotel area. In: Karst Hydrogeology and Karst Environment Protection (Proc. 21st Congress of IAH, Guilin, China, October 1988), vol. 2, 983–991. IAHS Publ. no. 176.

Özgür, C., Uzel, Ö. F. & Tamgaç. Ö. F. (1987). Kırşehir-Terme kaplıcası Terme 2 ve Terme 3 sıcaksu sondajları kuyu bitirme raporu. MTA Report no. 8222, Ankara (in Turkish).

Özmutaf, M. & Yüce, G. (1989). Yozgat-Boğazlıyan-Bahariye (Çavlak) kaplıcası sıcaksu sondajları (BB- I, BB-2) kuyu bitirme ve korunma alanlan etüdü raporu. MTA Report no. 8216, Ankara (in Turkish).

Quilong, Y., Kuide, X. & Zhang Zbenguo. (1985). Preliminary assessment of the geothermal resources of China. In: Jnr. Symp. on Geothermal Energy (Hawaii, USA), 43–52.

Rybach, L. & Hauber, L. (1990). Swiss geothermal energy update 1985–1990. Geothermal Resources Council Transactions, vol. 14, Part l, 239–246, Hawaii, USA.

Shaterev, K. D. (1976). Les acrotothermis de la peninsule Balk:anique. In: Proceedings of the Int. Congress on Thermal Waters, Geothermal Energy and Volcanism of the Mediterranean, vol. 2, 519–533, Athens.

Şahinci, A. (1988). Çeşme yöresindeki kaplıcaların hidrojeolojik - jeokimyasal özellikleri ve soğuk sularla ilişkileri. Ulusal I. Hidrojeoloji Sempozyomu. Bildiriler Kitabı, s. 103–109, Ankara (in Turkish).

Şimşek, Ş. (1985). Present status and future development of the Denizli-Kızıldere geothermal field of Turkey. In: International Symposium on Geothermal Energy (Hawaii, USA), 203–210.

Vandelberghe, N. (1990). Geothermal energy in Belgium: an overview Geothermal Resources Council Transactions, vol. 14, Part I, 9–14, Hawaii, USA.

Zhang Zhenguo. (1988). An assessment of karst geothermal resources of the North China basin. In: Karst Hydrogeology and Karst Environment Protection (Proc. 21st Congress of IAH, Guilin, China, October 1988), vol. 2, 968–974. IAHS Publ. No: 176.

The manufacturer's authorised representative in the EU is Springer Nature Customer Service Centre GmbH, Europaplatz 3, 69115 Heidelberg, Germany. If you have any concerns regarding our products, please contact ProductSafety@springernature.com

Printed and bound by CPI Group (UK) Ltd, Croydon, CR0 4YY
28/09/2026
02224097-0004